CW01656139

'WE PLOUGH THE FIELDS AND SCATTER'

THE HISTORY OF THE PETERBOROUGH AGRICULTURAL SOCIETY

Dedication

The Peterborough Agricultural Society played a very significant and enjoyable part in the lives of the Odam Family, in particular my father Frank, myself, my uncle Walter and his sons, Jack and Lawrence. It is for this reason that I commissioned the authors to write the history of the Peterborough Agricultural Society.

Joseph Odam July 1998.

WE PLOUGH THE FIELDS AND SCATTER

THE HISTORY OF THE PETERBOROUGH AGRICULTURAL SOCIETY

by
John Knight and David Brandon

Foreword by
H.R.H. The Prince Philip, Duke of Edinburgh, KG. KT.

Shaping out each old tradition-trodden track
Into furrows, fertile furrows, rich and black.
Oh, what harvests they will yield
When they widen to a field.
(Ella Wheeler Wilcox 1850-1919)

Published by
The East of England Agricultural Society
1999

First published in the United Kingdom in 1999 by
East of England Agricultural Society
East of England Showground, Peterborough, PE2 6XE

ISBN 0 9535629 0 5

All proceeds from the sale of this book will go to
the East of England Agricultural Society

Cover Illustration
Artist's impression of proposed 1880s development by Peterborough Land Company of the Park Road and Broadway areas of the City with high-class villas. Little of this development actually took place because the Company ran into financial difficulties. The picture embraces the area of the old Boroughbury Showground. Little's Farm can clearly be seen and this is one of the very few pictures of the period which shows the Fitzwilliam Hall with its domed roof, bottom right.

Produced for the Society by Past Historic, Kings Stanley, Gloucestershire
Printed and bound in Great Britain by MPG Books Ltd, Bodmin, Cornwall

BUCKINGHAM PALACE.

Throughout the history of the British Isles, the evolution of agriculture has closely mirrored social development. In the early days it was agricultural practices that largely determined the pattern of life in the countryside. All that changed with the industrial revolution and the continental wars of the 18th century. Subsistence agriculture virtually disappeared as the greatly increased urban population needed to be fed. The pressure on farming to produce large quantities of cheap food and to compensate for the difficulty of importing food during the wars was a very considerable incentive to improve agricultural techniques.

The Royal Society and the Royal Society of Arts had been taking an interest in the development of agriculture for a number of years before the first agricultural societies were formed by the farming and land-owning community. The Peterborough Agricultural Society was among the first to be formed with the object of assisting farmers to become more efficient and to make better use of the new technologies that were being developed by the urban industries.

Things changed yet again when the rural communities gradually declined as farming became less labour-intensive and with the attraction of industrial work. The balance had begun to swing away from the domination by the country over the city and it gradually dawned on the agricultural societies that there was a need to keep the city-dwellers informed about the country's agricultural industry. This vital 'public relations' function has now become just as important for agricultural societies as the original need to help farmers improve their techniques. By the end of the 20th century the pendulum will have swung decisively towards the city-dwellers, who will completely dominate what is left of the rural agricultural community.

The great value of this book is that it traces these changes through the evolution of the Peterborough, and then the East of England Agricultural Societies. It is a significant contribution to the social history of this country.

Philip

Contents

List of Illustrations

Pen Portraits

John Knight, M.Ed., is a well-known figure in Local History studies in and around the Peterborough area. He wrote and presented material for the Peterborough Evening Telegraph for six years providing the photographs and illustrations for their Local History features entitled, 'Then and Now', 'Nostalgia', and finally the larger format publication, 'The Peterborough Story', which once again used his illustrations. He was a professional engineer and then successively a lecturer, manager and trainer in Adult Education. He is a national speaker on a variety of historical subjects.

David Brandon, M.A., spent 20 years in Further Education finishing as a Business Manager before taking early retirement. He is now a visiting lecturer on Social and Economic History at eight Universities, as well as several Adult Residential Colleges and the Workers Educational Association. He also leads historically-based study tours. He has written many articles on the history of the Labour Movement.

In 1997 David Brandon set up, with John Knight, a partnership called 'Historical Projects' to research, write and publish Local and Social History material and to undertake commissions to write family and company histories.

Acknowledgements

The authors would like to thank the following for their generous help:

R.W. Bird and T. Gibson.

Also Messrs M. Baunton; M. Bibby; Mrs L.C. Brandon; Elizabeth Davies (Peterborough Museum & Art Gallery); E.A. Davies; P.G. Fryer; Sir Richard Greenbury; Messrs F.J. Grounds; S. Harris; A. Hemmant; R. Hillier (Peterborough City Library); J. R. Horrell; E. Hurn; M.W. Jeffreys; Mrs. H.M.T. Jones; G.P.L. Little; Rita McKenzie; Mrs J. Miller; Sir David Naish; Sir Philip Naylor- Leyland; Messrs J. Odam; W. Proby; Mrs J. Proby; Messrs W.R. Riddington; T. Robinson; G. Ruston; W. Samuel; J. N. Scorer; Mrs Janet Smart; Messrs J.M. Stansfield; G.T. Ward; R. Waterworth; R.A. Whitsed; A. Williams; B. Williams; additionally the Cambridgeshire Heritage Record Office and many others who have given time, encouragement and historical material but who have indicated they do not wish to be acknowledged.

Picture Credits

The authors gratefully acknowledge the following for permission to reproduce photographs and illustrations as follows:

P. and G.H. Warren: cover; H.R.H. Prince Philip, Duke of Edinburgh, foreword; Sir Philip Naylor-Leyland: 1; Mrs J. Proby: 2; F. Grounds; 3; Peterborough City Library: 4, 6, 7, 15; Peterborough City Museum: 16, 17, 20; R. McKenzie: 5, 22; J. Knight: 8, 9, 12, 13, 14, 26; and many others who have graciously contributed but do not seek acknowledgement.

CHAPTER ONE

Introduction

How many people who live in the Peterborough area today are aware that a host of devices that became or are still commonplace, such as the telegram, the telephone, the motor car and the aeroplane, made some of their earliest appearances in the City at the Peterborough Agricultural Society's Annual Shows? These Shows were always, as we would say today, at the leading edge when it came to demonstrating the practical uses of new forms of technology. It speaks volumes about the wider significance of the Society in the development of the region's economy that the innovations which appeared there were by no means exclusively concerned with agriculture.

The Society can proudly claim to have been a major and respected factor in the life of the City and its surrounding region over the two centuries that have elapsed since its founding in 1797. It was established at a time when the country was engaged in bitter hostilities with post-revolutionary France and when Bonaparte was emerging to start his climb to power. At the same time the growth of manufacturing industry, of the towns and of the urban population were making it urgent to find ways of increasing the output and productivity of agriculture.

The early years of the Society are somewhat shrouded in mystery and provide problems for the historian in that few reliable records are available before the late 1850s. We do not know where the early livestock shows held after the Society's initial inauguration were located and we do not even know whether its activities were continuous from 1803 to 1837. However, with the official re-inauguration of the Society in the late 1830s, it appears unlikely that they would have been. The authors' research has not been able to establish with absolute certainty where any shows were held prior to the establishment of a permanent site in a yard at the back of the Waggon and Horses public house in 1848. This yard consisted of about four acres of land and was located at the rear of where Woolworth's store now stands on Bridge Street.

Subsequent show locations are better documented but confusion can be caused by the Society's habit of staging events additional to the main Show at other times of the year. In 1874, for example, a Horse Show was held on 4 April, an Agricultural Engine Driving Trial on 20 May and a Sheep-Shearing Competition on 8 July. On 23 June in the same year a Wool Fair was held at the Cattle Market and a Ploughing Competition held on 1 October on farmland some distance from the City.

The Society's early years and its growth and stability were greatly assisted by the active presence of the 4th Earl Fitzwilliam who was elected President at the first inaugural meeting and who continued to serve in that capacity until his death in 1833. Members of the Fitzwilliam family have continued to lend their name and their selfless efforts to the work of the Society ever since. The endeavours of the Fitzwilliams, the sterling work of a host of unpaid and untiring Council and Committee members and the fine contribution made by the succession of hardworking and able Secretaries that the Society has appointed, helped to ensure that, even before its merger with other societies to form the East of England Agricultural Society, the Peterborough Society's prestige and influence was such that its Annual Show and other events were of far more than local significance.

An important element in the success of the Peterborough Agricultural Society has without doubt been its willingness to work closely with a variety of other partners to advance the interests of agriculture and of the area's rural community as a whole. For example, in 1865, 1871 and 1881, it went into partnership with the Northampton Agricultural Society in the staging of joint shows. These were sited firstly at Boroughbury, then on what used to be called Thorney Road, later Eastfield Road, adjacent to where the old cemetery now stands, and finally on 'The Rink' Showground and Little's Farm in what became the Park Road and Broadway area of the City.

Another example of the mutual benefit of this kind of relationship is the forging of a close union between the Peterborough Agricultural Society and the Foxhound Show Society from its formation by the 6th Earl Fitzwilliam in 1878. After this date the 'Hound Show' came traditionally to be held on the second day of the Peterborough Agricultural Society's Annual Show. In 1934 His Majesty King George V, recognising the importance of the Foxhound Show, graciously gave permission for the Foxhound Society henceforth to style itself, 'The Peterborough Royal Foxhound Show Society'. This added considerably to the prestige of the three day Annual Show event as a whole and equally it provided the Peterborough Royal Foxhound Show with an absolutely ideal staging ground for its own highly-regarded individual events which have continued to be presented throughout the period of the two Societies' ongoing partnership.

Agricultural Societies, and the Peterborough Society was among the earliest dozen or so, first appeared in the eighteenth century and from the start set out to encourage good practice in their local farming communities and to disseminate new ideas. There were exhibitions and displays of modern farming machinery, implements and techniques as well as a wide variety of competitions, many offering premiums in the form of cash prizes. These premiums could be offered for meritorious innovations, for high quality stock or crops and to those workers who could demonstrate particular skill in farming operations. They were also sometimes awarded, perhaps somewhat patronisingly, to those workers who had brought up large families without recourse to charity or the assistance of the local Poor Law authorities.

It could be argued that the agricultural show, as it evolved in Victorian times, was the direct descendant of the hundreds of traditional local fairs and markets, many of them with charters going back centuries, which had played such a vital role for so long in the social and economic life of the countryside. The agricultural show provided a much more controlled environment than the old fairs which had often been accompanied by drunkenness and uncouth revelry that could easily get out of hand. Consequently, these fairs were coming under increasing pressure to be abolished because of the almost obsessive fear which haunted the minds of the Victorian upper and middle classes that any sizeable gathering of working people presented a potential threat to law and order.

The Shows in Peterborough have been supported, certainly over recent years by increasing numbers of visitors. For example, 33,549 for the first Show at Alwalton to 61,940 at the first East of England Show in 1968 and 160,000 in 1996 and they have been drawn not only by the farming exhibits and competitions but by an ever-widening range of other attractions. The prime purpose, however, has always been and continues to be the encouragement of agricultural improvement in the broadest sense. It is therefore entirely natural that major innovations in agricultural practice and important new types of machinery have often made their first appearance in the region at the Society's Annual Show. In terms of machinery, arguably the most revolutionary was the steam plough and this made its first appearance in the City at the 1858 Show.

This current historical work attempts to trace in outline the major events and landmarks in the history of the Peterborough Agricultural Society and wherever possible to emphasise its intimate links with the people, institutions and contemporary development of the City of Peterborough and the surrounding region. It also draws attention to the wider economic, political, social and technological factors which affected agriculture as a whole and farming in the Peterborough area in particular. It takes this approach because the Society did not operate in a vacuum and was forced to anticipate, work with and sometimes retrospectively and even reluctantly to respond to a host of considerations which were often beyond its immediate control. In that sense the history of the Society is the history of the region's agriculture over a period of two hundred years.

In demonstrating its continuing ability to move and adapt to changing demands and varied circumstances, or in some cases even to advance ahead of the times, the Peterborough Agricultural Society and its descendent, the current East of England Agricultural Society, can be proud of having made a significant contribution to agricultural development at a regional and a national level as well as forging for themselves an integral, honourable and vital part in the life and the history of the City of Peterborough and the surrounding area.

1. Rt. Hon. 4th Earl Fitzwilliam, 1748-1833.

The 4th Earl Fitzwilliam of Milton near Peterborough, the first President of the Peterborough Agricultural Society. He established the tradition followed by many of his descendants of supporting and playing a prominant role in the growth and prestige of the Society.

2. Rt. Hon. The Earl of Carysfort, 1751-1828.

With the Fitzwilliams, members of the Carysfort family of Elton near Peterborough provided much support for the Peterborough Agricultural Society in its early years.

CHAPTER TWO

The Early Days: 1797-1864

The Peterborough Agricultural Society was inaugurated at a well-publicised meeting held in the old-established Angel Hotel which used to stand in the City Centre, on 10 January, 1797, electing as its first President the 4th Earl Fitzwilliam. This momentous meeting was recorded in *The Lincoln, Rutland and Stamford Mercury* newspaper of Friday, 3 February, 1797.

The annual subscription was set at one guinea and remarkably remained at that amount until the 1950s. Twelve guineas entitled a gentleman to become a member for life. It was agreed that the Society would meet three times a year, on the Wednesday of Easter week, the first Wednesday in August and finally on the third Wednesday in December. A full list of the inaugural subscribers was issued. This provides a valuable historic record:

The Right Honourable Earl Fitzwilliam (President)
The Right Honourable the Earl of Carysfort

William Waller, Esq., Chesterton
Wt. Thomas Squire, Esq., Peterborough
William Squire, Esq., Peterborough
John Clare, Esq., London
Charles Cole, Esq., Southgate, Middlesex
Rogers Parker, Esq., Gt. Munden, Herts
Henry Cole, Esq., Spalding
George Maxwell, Esq., Spalding
John Wing, Esq., Thorney
John Brickwood, Esq., London
William Walcot, Esq., Oundle
Rev.C.B. Phillipson, Peterborough
Rev. John Girdlestone, Thorney
Rev. William Strong, Peterborough
Mr Francis Hopkinson, Peterborough
Mr Thomas Mewburn, Stanground
Mr William Bailey, Jun., Glinton
Mr Charles Davies, Chesterton
Mr Thomas Simpkin, Gunthorpe
Mr Wlliam Bowker, Peterborough
Mr George Edmunds, Peterborough
Mr John Bailey, Thorney
Mr Peter Ellis, Peterborough
Mr Charles Berkley, Biggin, Oundle
Mr Thomas Goodman, Peterborough
Mr Clement Whitehead, Werrington
Mr Isaac Pears, Thorney
Mr James Barber, Willow Hall
Mr William Freeman, Peterborough

The first elected Secretary was Richard Weston and he was given the task of drawing up a book containing the Rules, Orders and Resolutions of the Society. Perhaps this proved too onerous a task because, for reasons unknown, he soon resigned, to be replaced by Mr Newton Bosworth who received a salary of five guineas per annum. The rule book was finally produced and issued in the following year, 1798, its contents having been endorsed at the Society's AGM which had been held on Wednesday, 20 December, 1797. This document shows that the list of subscribers had risen to 40. The extra subscribers were:

Thos. Orby Hunter, Esq., Thorpe Hall
Mr William Waldgrave, Castor
Rev. Mr Loddington, Oundle
Mr William Platel, Peterborough
Arthur Mackay, Huntingdon
Mr Joseph Newton, Willow Hall
Mr John Hicks, Fotheringhay
Mr William Hardwick, Deeping
Captain Hart, Woodstone

The resolutions agreed by the Society at the time make interesting reading:

1. *That every premium which may be offered by this Society, for the Improvement of Agriculture, or for any other improvements, be confined within a circle of 20 miles from Peterborough, except, in such cases as the Society may find it necessary to offer their premiums to the public in general.*

2. *That a premium of ten guineas be offered to the public in general for any person who will communicate to the Society an effectual method of destroying the wire-worm in the land.*

3. *That the following premiums be offered for the year ensuing, to persons resident within twenty miles of Peterborough; viz.*

Five guineas to the owner of the best two-shear ram.

Three guineas to the owner of the best shearing ram.

Five guineas to the person who produces the best yearling draught stallion.

Five guineas to the person who produces the best bull, not more than two years old when shown.

Two guineas to the person who produces the best boar, not more than 18 months old, when shown.

N.B. No stock to be entitled to a premium, without full proof of its being bred by the claimant.

Two guineas to the labourer, in husbandry, who has brought up the most numerous family, without parochial assistance.

One guinea to the labourer who has worked for one master, or on one farm the longest time: he working for that master, or on that farm, at the time he makes his claim.

One guinea to the male servant who has lived the longest time with any one master or mistress in husbandry.

One guinea to the female servant who has lived the longest time with any one master or mistress in husbandry.

N.B. No claims will be admitted but for those who have lived, or worked, at least seven years in each of the above situations.

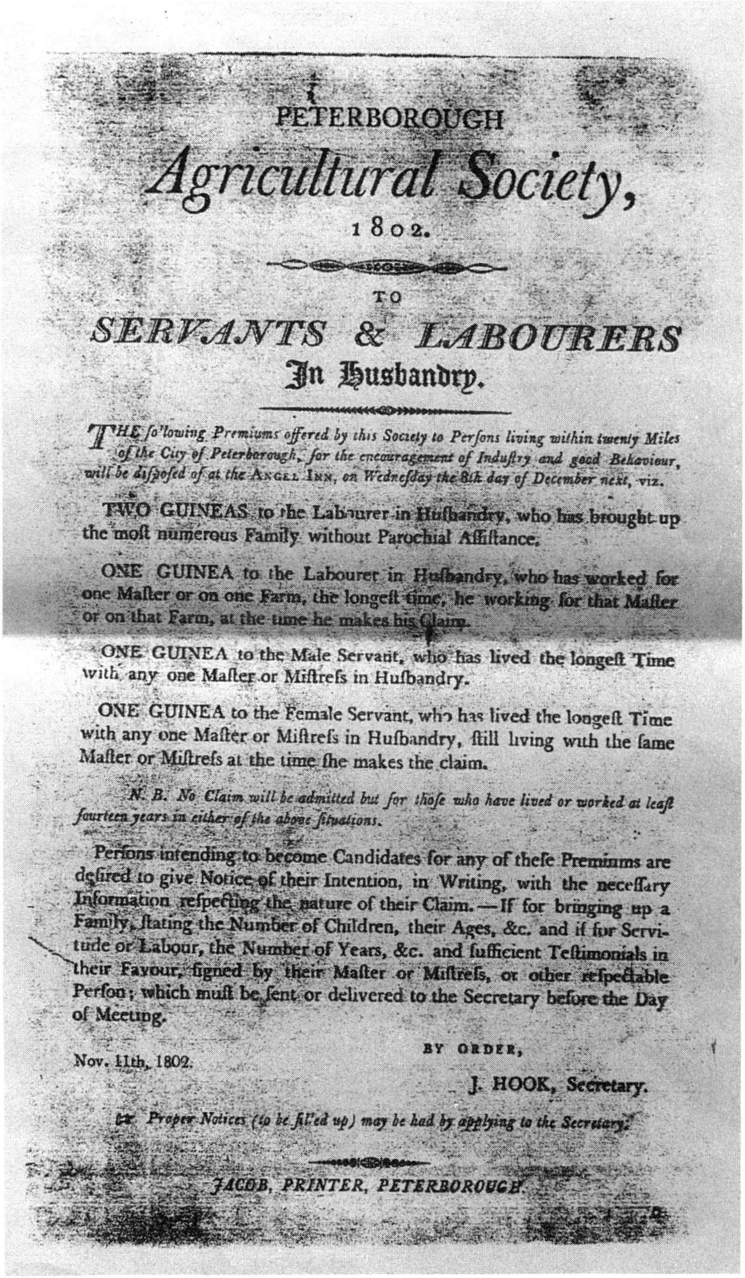

PETERBOROUGH

Agricultural Society,

1802.

TO

SERVANTS & LABOURERS

In Husbandry.

THE following Premiums offered by this Society to Perſons living within twenty Miles of the City of Peterborough, for the encouragement of Induſtry and good Behaviour, will be diſpoſed of at the ANGEL INN, *on Wedneſday the 8th day of December next, viz.*

TWO GUINEAS to the Labourer in Huſbandry, who has brought up the moſt numerous Family without Parochial Aſſiſtance.

ONE GUINEA to the Labourer in Huſbandry, who has worked for one Maſter or on one Farm, the longeſt time, he working for that Maſter or on that Farm, at the time he makes his Claim.

ONE GUINEA to the Male Servant, who has lived the longeſt Time with any one Maſter or Miſtreſs in Huſbandry.

ONE GUINEA to the Female Servant, who has lived the longeſt Time with any one Maſter or Miſtreſs in Huſbandry, ſtill living with the ſame Maſter or Miſtreſs at the time ſhe makes the claim.

N. B. No Claim will be admitted but for thoſe who have lived or worked at leaſt fourteen years in either of the above ſituations.

Perſons intending to become Candidates for any of theſe Premiums are deſired to give Notice of their Intention, in Writing, with the neceſſary Information reſpecting the nature of their Claim.—If for bringing up a Family, ſtating the Number of Children, their Ages, &c. and if for Servitude or Labour, the Number of Years, &c. and ſufficient Teſtimonials in their Favour, ſigned by their Maſter or Miſtreſs, or other reſpectable Perſon; which muſt be ſent or delivered to the Secretary before the Day of Meeting.

Nov. 11th, 1802.

BY ORDER,

J. HOOK, Secretary.

☞ *Proper Notices (to be fill'd up) may be had by applying to the Secretary.*

JACOB, PRINTER, PETERBOROUGH.

3. 'To encourage Diligence and Thrift'.

Handbill dated 1802 and published by the Peterborough Agricultural Society, inviting applicants for awards in the category 'Servants and Labourers in Husbandry'. This rare item relates to the activities of the Peterborough Agricultural Society soon after its initial inauguration in 1797.

It was also resolved that :

> *the premiums offered for horses be decided and disposed of at the Easter meeting, those for other cattle, at the August meeting; while premiums payable in respect of experiments for the improvements of grain and other seeds, as well as those for servants and labourers, be decided and disposed of at the annual meeting in December.*

The Early Background

The Peterborough Agricultural Society was formed at an extremely critical time for British agriculture, both in the long and the short-term. The country was then undergoing the process which historians have called 'The Industrial Revolution'. This involved an extremely complicated series of interacting processes, the effect of which can most easily be summarised by saying that, over a period of about a hundred to a hundred and fifty years, the country's economic and social structure was transformed from a largely rural and agricultural basis, to a predominantly urban one. In the nineteenth century, the English countryside was to see more rapid and radical change than had taken place perhaps for the ten previous centuries, although it has to be said that some of these changes involved the continuation or intensification of processes which had been occurring for many generations.

Among the most obvious change was the enclosure of open fields, commons, heaths and greens. Enclosures had actually been taking place randomly for centuries, but the pace and volume accelerated considerably in the eighteenth century and the process became far more systematic. It has been estimated that in the hundred years after 1750, about 6 million acres, a very substantial percentage of all agricultural land and centred particularly in the Midlands and the South East of the country, underwent this process.

The social effects of enclosure have long been debated by historians and it seems that the greatest beneficiaries, at least in the short term, were the landowners and large farmers but even the smaller tenant in some cases made useful gains. Enclosures often coincided with the abolition of tithes, with improvements to drainage, a widening use of mechanical aids and the application of new farming techniques, the building of new, more convenient roads in the parish and, overall, a substantial rise in the value of land. It is easy to make misleading generalisations about the effects of a process as complex as that of enclosure and in practice the results varied considerably from parish to parish, even within a region or district.

At this time, Peterborough was administratively within the County of Northamptonshire and the records show that this county was among those most profoundly affected by the process of enclosure, something which was by no means greeted with universal acclaim. Enclosure involved the creation of new fences and the laying of hedges, the stopping up of ancient meandering lanes and the bringing into cultivation of large areas of hitherto largely unused waste and heathland. Through the years, many people have expressed their delight at the characteristic hedged landscape of enclosures in the East Midlands but at the time not everybody was susceptible to this aesthetic charm. For example, John Clare (1793-1864), the 'Peasant Poet' from the village of Helpston, a few miles north of Peterborough, much preferred the existing wilder, more natural countryside to what he saw as an essentially imposed pattern of regular fields and orderly hedgerows and he gave vent to his indignation in verse:

Inclosure, thou'rt a curse upon the land,
And tasteless was the wretch who thy existence plann'd.

Large-scale reclamation of the Fens was also taking place at this time. To view the Fenland as it must have looked before drainage, almost the only place we can go today is Wicken Fen in Cambridgeshire but throughout the vast tract of Fenland that stretches east from Peterborough, there remain some traces of the complex methods that were used for water-removal. These ranged from early wind-driven scoop wheels, through a variety of steam and diesel pumps, to the electric pumps of today.

Paradoxically, agriculture's decline was only a relative one compared to industry's massive growth and the efforts of the nation's farmers were absolutely crucial in the process of feeding a greatly increased population, large numbers of whom were town-dwellers who obviously lacked the opportunity to grow any crops and had little space to rear any animals of their own. In return for their labour, they received wages with which they were able to buy bread and other staple items of diet which were produced for sale by the country's farming community.

Subsistence farming was very much in decline and agriculture now largely produced for a nation-wide market economy. The pressure on farmers to increase their output and productivity and to embrace practical innovations was very considerable. The Industrial Revolution was accompanied by a much greater appreciation of, and willingness to apply, theoretical and practical scientific and technological knowledge and principles. British agriculture was, by the standards of the time, uniquely innovative and progressive. It focused on supplying the growing need of the developing market for a range of cheap, home-grown agricultural produce. Britain, at the time, was the only country in Europe from which the peasantry had virtually disappeared. For all that, there was much need for progress. It would, for example, be no real exaggeration to say that the use of animal power was scarcely more effective in 1800 than it had been 500 years earlier in medieval times.

It was this spirit of enquiry and innovation which forms the background against which a number of national institutions were set up which were to take an interest in agriculture. These included the Royal Society (1660), the Society of Arts (1754) and the Royal Horticultural Society (1804). From the start, the second of these, which eventually became the Royal Society of Arts (RSA), offered prizes and medals for progressive innovations and its first award to the agricultural sector was in 1757 and concerned a set of experiments intended to make better use of natural fertiliser. By 1761 the RSA had established an Agricultural Committee which actively encouraged experiments and trials and gave awards where appropriate. Early awards were given to support the introduction and cultivation of new crops, for innovative, effective forms of animal husbandry and the design and utilisation of various new farming implements.

The first Agricultural Society in Britain appears to have been the Brecknockshire Society founded in 1755. There were 90 Societies by 1835 and over 700 by 1860. Probably the best known of the pioneer Societies was the Bath and West founded in 1777, which sponsored agricultural shows at sites throughout southern England. However, many early Societies were very small-scale and their activities seem to have consisted of not much more than regular discussion meetings and the development of a reference library for the use of the members.

In 1793 the Board of Agriculture was set up. It had financial support from the Government but, rather than giving the Board useful kudos, this official involvement met with some suspicion within the farming community, probably because they regarded it as the thin end of the wedge for possible Government intervention in the industry. The most enthusiastic voice on behalf of the Board of Agriculture was that of Sir John Sinclair, who owned large estates in Scotland and was M.P. for Caithness. He saw the Board's function as being one of spreading information about improved forms of farm management and diffusing up-to-date knowledge about new implements, equipment and techniques. He also saw the Board providing some direction and support for the numerous local agricultural societies which were becoming established at that time and which were, with varying degrees of success, seeking to disseminate knowledge of good practice and innovations among their members.

The Peterborough Society was established at a critical time because the country was in the throes of the Wars with France which were almost continuous between 1793 and 1815. These years had seen Napoleon Bonaparte emerge and begin his extraordinary drive for international domination. At one stage it looked as if the only real factor that could frustrate his ambitions was Britain's naval and military power. He attempted to cut Britain off from all its vital international trading relationships, hoping thereby to starve his stubborn and isolated enemy into submission. It is possible that the daily progress of the war was followed with more than usual interest in the Peterborough area because of the vast prisoner-of-war camp that was built at nearby Norman Cross. This had a beneficial impact on the local economy and gave local people the daily sight of prisoners-of-war from France and its allies, because many of the prisoners worked on farms in the neighbourhood.

Although Bonaparte proved unsuccessful in this effort, the Wars provided a useful boost for British agriculture because high prices further encouraged farmers to expand and develop. Additionally, exceptionally bad weather from 1795 to 1800 and 1808 to 1812 meant that, as a direct result, grain was in such short supply that prices rocketed. The process of enclosure accelerated, unused tracts of land were brought under the plough, the practice of crop rotation intensified and drainage schemes were extended. Greater use was made of artificial fertilisers and animal husbandry was also improved.

After the Wars there was a slump in prices, most of all among the wheat-growing farmers of Southern and Eastern England who were loud in their protestations to Parliament about hard times and unfair conditions. By contrast with the golden days of the war, times were indeed hard for these farmers but it should be borne in mind that they represented only one section of the industry. Mixed farmers were less adversely hit while livestock producers enjoyed a slow but rising demand for meat from the growing urban areas. While this was certainly not a good period for agriculture as a whole, paradoxically the conditions actually created an incentive for progressive farmers to innovate since the more efficient ones were often better placed to ride out the relatively depressed times.

Peterborough is situated literally astride the geological divide between the gently undulating upland country of Northamptonshire, Rutland and the western part of Huntingdonshire on the one hand, and, on the other, the flat and then still largely drowned Fenland immediately to the east. While drainage of the fens had been proceeding on a systematic basis since the 1630s, and sporadically before that, the great expanse of Whittlesey Mere, very close to Peterborough, was not drained until the early 1850s. Further east, Wisbech stood at the centre of a large stretch of at least partially drained and cultivated fen. Peterborough's location at the meeting point of two so very different topographical systems has always influenced its history. It had a considerable bearing on agricultural land use in the area and this has been reflected in the membership of the Society and in its operations. Speaking in very general terms, it can be said that the majority of its members have tended to come from the district east of Peterborough.

After the Wars with France, there were three main issues about which farmers felt particularly strongly. These were the necessity of maintaining protection for agriculture from cheap imports, the requirement to tackle inflation by returning to the gold standard, that is, the practice of making bank notes convertible to gold on demand and thirdly, the urgency of repealing the malt tax, which it was claimed was crippling barley growers. Agricultural Societies set themselves the laudable objective of being non party-political and the Peterborough Society we can assume was no exception. Clearly its individual members had their axes to grind and might make their views clear in speeches on such occasions as the Society's annual dinner but there is no evidence that the Society attempted to act at any time as a political lobbying body.

It was, perhaps, inevitable that agricultural societies tended to contain those members of the local farming community who were more open to the idea of research and experimentation as a vital contribution to the advance of agriculture and who could, for that reason, justifiably be described as 'progressive' farmers. They were often the leaders of the agricultural interest in their locality and so were naturally likely to be the people who established and managed these societies. At the same time, they were often more concerned with perceived social quality than quantity among their members and they would usually aim to have a well-known local landed aristocrat or member of the titled landed gentry as President, for example. The possession of land had a symbolic importance that reached far beyond its obvious economic and political significance. The rural land-owning class presided over virtually all aspects of country life at this time and enjoyed a real sense of hegemony based on aristocratic values and landed interest which continued to have a pre-eminent place long after the economic significance of agriculture in the life of the nation had gone into relative decline.

The list of early subscribers for the Peterborough Society is typical of these societies as a whole in containing a mixture of doctors, academics and clergymen as well as large and medium landowners and tenant farmers. A prominent MP or two added a touch of 'je ne sais quoi' to the list. As far as practising farmers are concerned, it is unlikely that the early subscription lists of the Peterborough Society included many farmers who tended less than 100 acres.

The Second Inaugural Meeting

The possession of a list of initial subscribers is valuable but it really only emphasises the difficulties that exist for historians because there is very little additional evidence that has survived about the early years of the Society. We know that the Society was first established in 1797 and that it held some livestock competitions in 1799, 1800, 1801 and 1802 but we cannot determine whether these were open to non-members or even with absolute certainty where they were held. It seems likely that the Society's activity then lapsed because *The Lincoln, Rutland and Stamford Mercury* states that on 23 July, 1838 a meeting was to be organised inviting the attendance of people interested in forming an Agricultural Association:

> *embracing every Parish and Place, any part of which is within 15 miles of Peterborough, for the great advancement of Agriculture, the encouragement of Breeding, with Exhibitions of Stock at such seasons and under such restrictions as may appear most conducive to profitable Husbandry, and for the encouragement of skill, industry, and good conduct in the Agricultural Labourer.*

This re-inaugural meeting was held at the Angel Inn in Peterborough and was presided over by the 5th Earl Fitzwilliam. During the year he had been very active in the negotiations surrounding the formation of what was to become the Royal Agricultural Society of England and it appears that this may have fired his enthusiasm and made him determined to revive his local but lapsed Agricultural Society.

We know that some of the other early Agricultural Societies were rather insular affairs at first, perhaps doing little more than circulating information to their members and offering premiums to those who, perhaps at the annual general meeting, could provide evidence of especially good practice. Gradually, it seems, the idea of an exhibition or show emerged where competitions could take place offering awards for excellence to representatives of the farming community as a

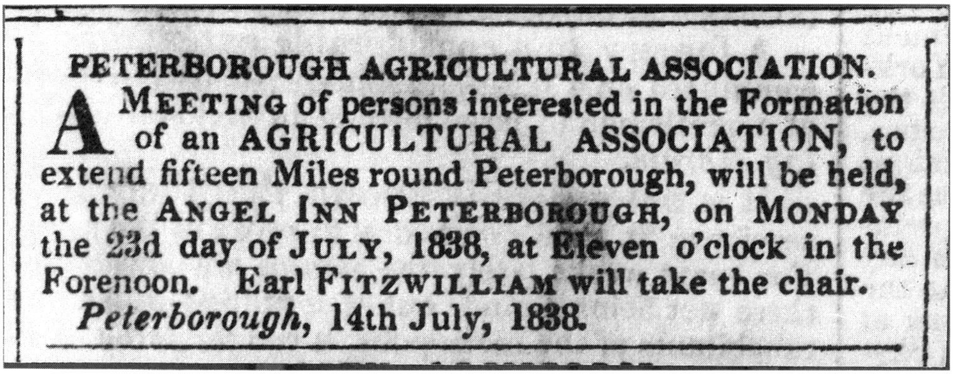

PETERBOROUGH AGRICULTURAL ASSOCIATION.

A MEETING of persons interested in the Formation of an AGRICULTURAL ASSOCIATION, to extend fifteen Miles round Peterborough, will be held, at the ANGEL INN PETERBOROUGH, on MONDAY the 23d day of JULY, 1838, at Eleven o'clock in the Forenoon. Earl FITZWILLIAM will take the chair.

Peterborough, 14th July, 1838.

4. Notice of meeting to re-form the Peterborough Agricultural Society.

Another scarce item from *The Lincoln, Rutland and Stamford Mercury* announcing a meeting on 23 July 1838 to attempt to re-form the Peterborough Agricultural Society.

whole. It was but a small step to throw the proceedings open to the general public and to charge them for the privilege. It is hardly surprising that the public then began to expect that in addition to the strictly agricultural items on offer, there should be additional attractions that were more in the nature of entertainment for the substantial numbers of people who passed through the gates but perhaps had little or no interest in the husbandry of the land.

A Society catalogue for 1840 exists indicating seasonal shows were to be held on: 4 March, 1 and 3 October-Labourers and Others Engaged in Husbandry; 8 April-Cattle and Horses; 11 July-Shepherds; 3 October-Cattle, Sheep, Fatstock, Pigs, Turnips and Implements. This catalogue clearly states that the Annual Shews (sic) will be held in the 'Wool Fair Yard' in Bridge Street, Peterborough. While this conceivably could be the same yard designated in the records as the 'Waggon and Horses Yard', it cannot be confirmed from historical records.

The Society certainly acquired a site of about 4 acres which it used for its annual shows at what was known as the 'Waggon and Horses Yard'. This was behind the public house of that name which formerly stood in Narrow Bridge Street, the yard being where the present British Telecom car park is located, at the rear of the Woolworth store in Bridge Street. This Showground, ideally situated in the City Centre, was in use from 1848 to 1864 and possibly earlier as shows of some kind were held by the Society during the period 1838 to 1847.

The Society's Shows after the Re-inauguration.

From the original formation of the Society in 1797 until the establishment of Peterborough's first successful newspaper, the *Peterborough Advertiser*, in May 1854, the majority of published information about its activities is in the form of snippets and sometimes larger items contained in *The Lincoln, Rutland and*

5. Waggon and Horses Public House, Narrow Street, Peterborough.

It was in the yard behind the handsome tiled facade of the Waggon & Horses, now demolished, that the first recorded shows of the Peterborough Agricultural Society took place. This pub stood close to the present-day junction of Bridge Street and Bourges Boulevard.

Stamford Mercury. Other information about events in Peterborough which sometimes includes snippets related to the Society, can be found in newspapers published in Bedfordshire, Cambridgeshire, Lincolnshire and Rutland with some more substantial pieces being found in the Huntingdonshire papers. We have to wait for several years until the *Peterborough Advertiser* came into production for the first detailed report of one of the Society's shows to appear and this can be found in the edition of Saturday 27 October, 1855.

It was reported that on the previous Wednesday the Society had held its annual exhibition of livestock and of farming implements. Some excellent beasts had been shown in addition to a range of first-rate implements but it was perhaps the contest for the best plough which had produced the keenest competition. Many of the implements on view were made by local firms and this provided them with an excellent marketing opportunity. These exhibitors in turn provided some financial support for the Show and helped to widen its drawing-power, particularly to the farming community.

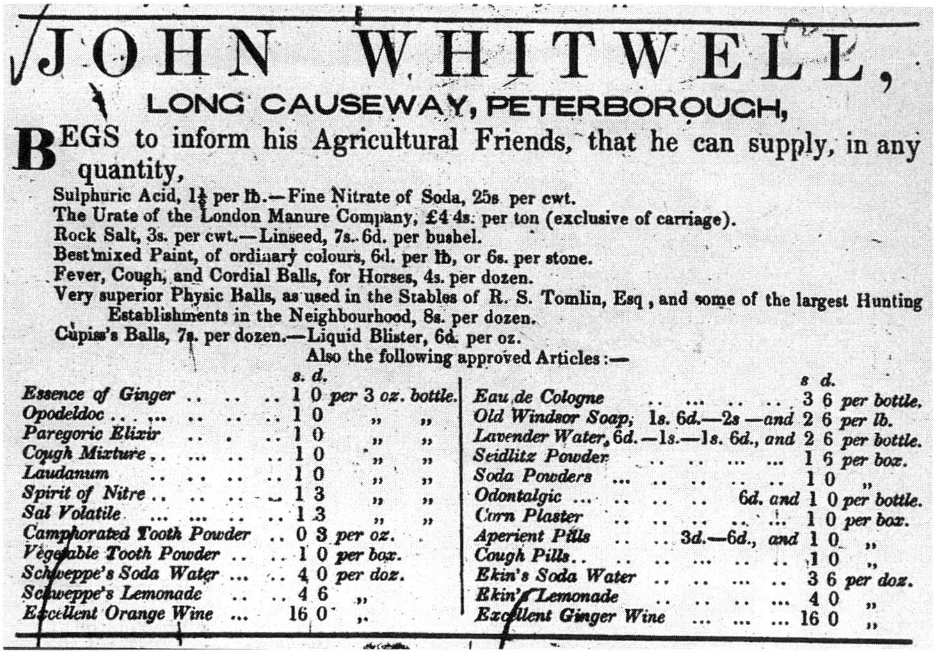

JOHN WHITWELL,

LONG CAUSEWAY, PETERBOROUGH,

BEGS to inform his Agricultural Friends, that he can supply, in any quantity,

Sulphuric Acid, 1¼ per lb.—Fine Nitrate of Soda, 25s per cwt.
The Urate of the London Manure Company, £4 4s. per ton (exclusive of carriage).
Rock Salt, 3s. per cwt.—Linseed, 7s. 6d. per bushel.
Best mixed Paint, of ordinary colours, 6d. per lb, or 6s. per stone.
Fever, Cough, and Cordial Balls, for Horses, 4s. per dozen.
Very superior Physic Balls, as used in the Stables of R. S. Tomlin, Esq, and some of the largest Hunting Establishments in the Neighbourhood, 8s. per dozen.
Cupiss's Balls, 7s. per dozen.—Liquid Blister, 6d. per oz.

Also the following approved Articles:—

	s. d.			s d.	
Essence of Ginger	1 0	*per 3 oz. bottle.*	*Eau de Cologne*	3 6	*per bottle.*
Opodeldoc	1 0	„ „	*Old Windsor Soap;* 1s. 6d.—2s —and	2 6	*per lb.*
Paregoric Elixir	1 0	„ „	*Lavender Water,* 6d.—1s.—1s. 6d., and	2 6	*per bottle.*
Cough Mixture	1 0	„ „	*Seidlitz Powder*	1 6	*per box.*
Laudanum	1 0	„ „	*Soda Powders*	1 0	„
Spirit of Nitre	1 3	„ „	*Odontalgic* ... . 6d. and	1 0	*per bottle.*
Sal Volatile	1 3	„ „	*Corn Plaster*	1 0	*per box.*
Camphorated Tooth Powder ..	0 3	*per oz.*	*Aperient Pills* .. .. 3d.—6d., and	1 0	„
Vegetable Tooth Powder	1 0	*per box.*	*Cough Pills*	1 0	„
Schweppe's Soda Water	4 0	*per doz.*	*Ekin's Soda Water*	3 6	*per doz.*
Schweppe's Lemonade	4 6	„	*Ekin's Lemonade*	4 0	„
Excellent Orange Wine ...	16 0	„	*Excellent Ginger Wine*	16 0	„

6. Advertisment for J. Whitwell.

Published in the *Peterborough Advertiser*, April 1846, this advertisement for the services of J. Whitwell of Long Causeway provides evidence that at this late date there were still businesses very similar to apothecaries, quacks and horse doctors who had traditionally played a major role in human and animal healthcare. In fact Mr Whitwell seems to have tried to be all things to everybody by selling soft drinks and toiletries as well.

Among the exhibitors was Mr W.P. Stanley who unquestionably made a major contribution to Peterborough's development as a centre of the engineering industry. In 1846 he was advertising his patent compound lever hand thrashing machines, guillotine chaff engines and iron harrows. A local man, starting out as an ironmonger, Stanley went on in 1853 to buy the erstwhile Buckle's brewery on land in Queen Street where the Queensgate shopping centre now stands. This he converted into an ironworks where he established a reputation for the manufacture and marketing of high-quality agricultural implements, luckily doing so at a time when there was a growing demand from the farming community for efficient labour-saving machinery. Stanley is a figure of considerable importance in Peterborough's history because his activities helped to establish the City as a centre for high quality precision engineering. Its reputation has continued to this day. His memory lives on in the name of the Stanley Recreation Ground which is close to the City Centre and adjacent to Brook Street. He was a well-known exhibitor at the Society's Shows and in 1855 he entered a portable

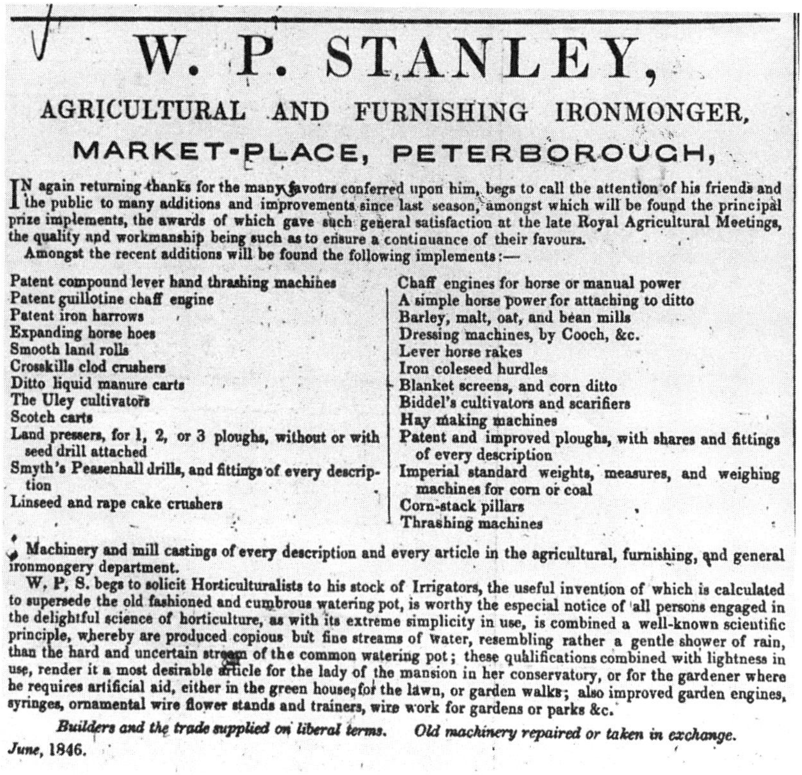

W. P. STANLEY,

AGRICULTURAL AND FURNISHING IRONMONGER,

MARKET-PLACE, PETERBOROUGH,

IN again returning thanks for the many favours conferred upon him, begs to call the attention of his friends and the public to many additions and improvements since last season, amongst which will be found the principal prize implements, the awards of which gave such general satisfaction at the late Royal Agricultural Meetings, the quality and workmanship being such as to ensure a continuance of their favours.

Amongst the recent additions will be found the following implements:—

Patent compound lever hand thrashing machines
Patent guillotine chaff engine
Patent iron harrows
Expanding horse hoes
Smooth land rolls
Crosskills clod crushers
Ditto liquid manure carts
The Uley cultivators
Scotch carts
Land pressers, for 1, 2, or 3 ploughs, without or with seed drill attached
Smyth's Peasenhall drills, and fittings of every description
Linseed and rape cake crushers
Chaff engines for horse or manual power
A simple horse power for attaching to ditto
Barley, malt, oat, and bean mills
Dressing machines, by Cooch, &c.
Lever horse rakes
Iron coleseed hurdles
Blanket screens, and corn ditto
Biddel's cultivators and scarifiers
Hay making machines
Patent and improved ploughs, with shares and fittings of every description
Imperial standard weights, measures, and weighing machines for corn or coal
Corn-stack pillars
Thrashing machines

Machinery and mill castings of every description and every article in the agricultural, furnishing, and general ironmongery department.

W. P. S. begs to solicit Horticulturalists to his stock of Irrigators, the useful invention of which is calculated to supersede the old fashioned and cumbrous watering pot, is worthy the especial notice of all persons engaged in the delightful science of horticulture, as with its extreme simplicity in use, is combined a well-known scientific principle, whereby are produced copious but fine streams of water, resembling rather a gentle shower of rain, than the hard and uncertain stream of the common watering pot; these qualifications combined with lightness in use, render it a most desirable article for the lady of the mansion in her conservatory, or for the gardener where he requires artificial aid, either in the green house, for the lawn, or garden walks; also improved garden engines, syringes, ornamental wire flower stands and trainers, wire work for gardens or parks &c.

Builders and the trade supplied on liberal terms. Old machinery repaired or taken in exchange.

June, 1846.

7. Advertisement for W.P. Stanley.

The *Peterborough Advertiser* in 1846 carried this advertisement for W.P. Stanley, Agricultural and Furnishing Ironmonger, soliciting custom for the Annual Show of that year. Historically, Mr. Stanley played a prominent role in the affairs of the City in the middle of the nineteenth century and his name is commemorated in the 'Stanley Recreation Ground' in Crawthorne Road.

steam engine and an improved chaff-cutting machine complete with a safety device that not only stopped the machine but also reversed the rollers in the event of an accident. Stanley was also a regular and successful exhibitor at The Royal Agricultural Society's Shows and indeed he was awarded a prize at eight consecutive 'Royal' Shows.

Although Stanley died in 1858, the company he founded at the Queen Street Ironworks, later became Amies and Barford, one of the country's leading manufacturers, particularly of rollers but also of general agricultural and horticultural equipment. Amies and Barford in turn evolved into Barford and Perkins who became prominent in the manufacture of steam ploughing machinery and later, under the guise of Aveling-Barford, were to become a large-scale manufacturer of steam and motorised road rollers although their operations were not carried out in Peterborough.

Much later, in 1932, J.E.S. and F.A. Perkins moved into one part of the site and started to manufacture diesel engines. Their present-day successor is known as Perkins Engines Limited, part of the global Caterpillar Group and is still a major employer in the city, its factory being on the eastern side of the City. However, it vacated the Queen Street site in 1975. This prestigious company can therefore trace its origins directly back to the Queen Street factory established by W.P. Stanley over one hundred and fifty years ago.

Mr G. Vergette also exhibited some ingenious implements at the 1855 Show including an improved chaff-cutter with the operating lever placed so that body pressure, rather than manual action would stop or reverse the machine. In addition, Mr S. Hand, of Glinton Iron Works exhibited his improved corn-mills and other articles and he won a prize for his horse-hoe. However, the farming implement which apparently attracted most attention from the judges and which secured a prize as the 'best improved agricultural machine not in ordinary use in the neighbourhood' was a dressing and weighing machine, made and exhibited by Mr Thomas Luck of Spalding Common. This in a simple and effective manner dressed and weighed corn at the same time. As soon as there was the required amount of corn in the sack, down went the scale, stopping the supply, thus preventing the possibility of an error in the weight.

At this time, the agricultural machinery and implement industry was very much in a state of transition. As early forms of mechanical aids came into use along with improvements to existing implements, so the industry began to change from what previously had almost been a 'cottage industry'. Here large numbers of village blacksmiths and others at the humble end of the rural hierarchy were prepared to turn their hands to addressing practical problems and to designing a host of devices intended to boost productivity on the land. Some were glorious failures and others may have been more successful and perhaps made a localised impact. However, the tendency was increasingly towards greater capitalisation of the industry and production largely went over to a factory basis which allowed larger companies to enjoy such benefits as economies of scale and more sophisticated marketing techniques. A number of towns in the eastern counties, such as Lincoln and Ipswich and to a lesser extent Peterborough and Grantham gained reputations as centres for the manufacture of agricultural machinery. The numerous companies based in such places were able to take advantage of the growing railway network to transport their products to agricultural shows throughout the country and of course, to their customers everywhere.

The railways played a crucial part in assisting the development of the more important county and local shows and Peterborough, which had become a major railway centre by the 1860s, was better situated than most to take advantage of the services they could provide. Railways facilitated the safe, quick and economical transport to the Show of both livestock and implements and also provided the means whereby the public could travel to Shows from outside their immediate area. The Great Northern Railway was quick to offer cheap passenger excursion tickets to Peterborough for the duration of the Show from nearby places such as Huntingdon, Biggleswade, Spalding, Boston and Grantham as well as a number of the smaller wayside stations on the lines serving those towns. The other railway companies which had reached Peterborough around this time offered similar facilities, although perhaps with rather less conviction.

In 1856 the livestock exhibited at the Show was described as being of a very high quality and the prize ox and prize bull were unanimously adjudged to be quite outstanding examples of their class, but unfortunately the public attendance was poor. At its usual after-show dinner at the Angel Hotel, a proposal was made that the Society should invite the Royal Agricultural Society of England to visit Peterborough and hold its Annual Show in the City. The members agreed, firmly believing that a visit from the 'Royal' would be of benefit to the Society. The rather quaint statement was made, that the Society should: *Be not the first by whom the new is tried, nor yet the last to lay the old aside.*

The Show of 1857 was rather a mediocre one, except in one respect and was undoubtedly overshadowed by the death, shortly before, of the President, the 5th Earl Fitzwilliam. The livestock on show was described by the local paper as 'middling' and the implements displayed no better than 'poor'. However a considerable impact was made in the competition for the best agricultural implement not then in general use. This was won by Mr Hayes of Elton, for his straw elevator, first patented in 1853.

The Straw Elevator, designed and patented by James Hayes of Elton, as first used in 1853.

8. Early Straw Elevator.

An engraving of a machine designed and patented by James Hayes of Elton near Peterborough. This was first used in 1853 and was exhibited at a number of Peterborough and other shows in the following years.

James Hayes was born in 1799 at the village of Elton, west of Peterborough and died nearby at Warmington at the ripe old age of ninety-two. He was the inventor and patentee of this extremely reliable elevator which could effect considerable savings in labour. All his machines had the name 'James Hayes, Overend, Elton' applied to them and so his name became well-known throughout the corn-growing areas of Britain at this time. The first trials of his elevator were held at Morborne, near Elton in 1853, but were not wholly successful. However, subsequent modifications turned it into a highly practical machine for which Hayes wisely decided to take out a patent. Complete with this patent, he then entered into a very profitable working agreement with three major national companies, namely: Clayton and Shuttleworth of Lincoln; Wallace and Stevens of Basingstoke and Charles Burrell of Thetford which gave them the right to manufacture and market these machines in return for which they paid him a royalty and also affixed to each machine a plate endorsing him as the patentee.

British Agriculture in Mid-Century

Sir Robert Peel had introduced the bill which repealed the controversial Corn Laws in 1846 and some historians have argued that this was part of a deliberate ploy to encourage the farming community to embark with more determination on cost-cutting exercises by apparently opening them up to the threat of increased competition. In the same speech as that in which he proposed repeal, he went on to outline a series of measures for supporting drainage works and encouraging increased productivity. Whatever Peel's intentions, the mid-1840s were undoubtedly the start of a period of widespread modernisation in British agriculture, sometimes referred to as 'High Farming' or the 'Golden Age of British Agriculture'. More will be said about 'High Farming' later in this chapter.

Even under the Corn Laws there had been a considerable increase in publications about agriculture and this provided clear evidence of the growing eagerness of the farming community to consider new ideas, processes and techniques. The *Cyclopaedia of Agriculture* appeared in 1825 and the *Quarterly Journal of Agriculture* came three years later. In the 1830s the *Mark Lane Express* and *Farmers Magazine* started publication and these were followed by the *Agricultural Express* in the 1840s. A crucial pioneering work on fertilisers, von Leibig's *Organic Chemistry in its applications to Agriculture and Physiology* appeared in 1840 and this was followed by many additional textbooks, including the influential *Elements of Practical Agriculture* by David Low.

Across the country, farmers' clubs, agricultural associations and scientific and educational institutions with a bearing on farming were appearing and expanding. Symbolic of this thirst for new ideas were the establishment by Sir John Laws of an agricultural research station at Rothamsted in Hertfordshire in 1843 and in 1846 the opening of the Royal Agricultural College at Cirencester.

Shortly before, in 1838, the Agricultural Society of England had been inaugurated, gaining the 'Royal' in its title in 1840. Its purpose was to link manufacturers and scientists on the one hand, with farmers and landowners on the other, to circulate relevant information, to correspond with other societies, to encourage experimentation, to give prizes for deserving scientific developments, to support relevant educational developments, to encourage improvements in veterinary surgery and to provide prizes for examples of exceptional husbandry and animal rearing. Similarly to the local Agricultural Societies, the Royal declared itself non-political from its inception, maintaining a discreet official silence during the heated debates in Parliament over the Corn Laws, but it is significant that some at least of its leading figures thought the way forward for British agriculture lay in improvement and self-help rather than protection. Significantly, this important body of opinion included the 5th Earl Fitzwilliam and Earl Spencer, the former in particular being immensely influential in both local and national politics as well as in agricultural circles. The Royal Society may not have been overtly political but its immediate ancestor, the short-lived Central Agricultural Society set up in 1835 was designed to help safeguard the traditional dominance of the landed aristocracy in the affairs of the nation. The 'Royal' itself was definitely the creature of the great landlords and richer tenant farmers.

Just as the country's manufacturing industry was developing along capitalist lines, so 'High Farming' was the application of capitalist, profit-seeking motivation to agricultural production. In that sense, 'High Farming' was the forerunner of today's high-input and high-output agricultural practice. The enterprising arable farmers of this period, especially on the claylands, were those who swiftly availed themselves of improved but cheaper drainpipes and tiles, thereby making their land easier to work. They then added natural fertilisers, such as guano which mostly originated from Peru and consisted of the droppings of seabirds. Wondrous claims were being made for the benefits of this substance and for a range of superphosphate products that were appearing on the market at this time and which enabled farmers to enjoy considerably enhanced yields. They were also likely to turn their attention to the increasing range of mechanical aids that were coming onto the market.

It could be argued that in 1800 the use of animal power was scarcely more effective than it had been in medieval times but certainly by the 1830s and 1840s, technological advances in iron manufacturing and in mechanical engineering especially with the manufacture of steam engines, was opening up a rich seam of possibilities for enhancing output, productivity and, it was hoped, increased profits. So farming was moving into the age of technology. First of all improved cultivating equipment appeared including rollers, clod crushers and far more durable ploughs and, in the late 1840s, numerous steam-powered devices for thrashing as it was then usually called, for winnowing and such purposes as hay and chaff-cutting, root-slicing and driving the machinery for bone-mills and guano-crushers. Potentially the most revolutionary and far-reaching use of steam power on the farm was in its application to the process of ploughing. This was first demonstrated in the Peterborough area by Lord Willoughby de Eresby at Edenham near Bourne in 1850.

It is to John Fowler of Leeds that most credit must go for making a practical success of ploughing by steam. In 1858 the Royal Agricultural Society awarded Fowler a prize of £500 for his steam-ploughing set. Two years later a great improvement was made by the addition of a second engine on the other side of the field in place of the anchor. With this system, each engine in turn drew the plough to itself and then moved along the headland to the width of the plough to prepare for the next furrow. This allowed six usually, or as many as eight, furrows to be ploughed at a time.

Few but the largest farmers could afford to buy their own steam-ploughing equipment and so the work was usually carried out by contractors with their own team of men, much as happens in the modern day with the use of combines and other farming machinery. The demand for contractors' services grew rapidly, for they enabled the farmer on heavy land to break up and make better use of soil which had previously been too hard for horses. Where the land was much lighter, the use of the steam-ploughing contractor left the farmer free to use his teams of horses for other work.

It would be misleading to suggest that everything at that time was plain sailing because there were a number of teething problems in the early days of steam ploughing. Ideally for steam ploughing purposes, fields would have been in the shape of perfect rectangles but unfortunately these were rarely to be found and the engine drivers had to wrestle with all the perplexities created by such hazards as irregular field boundaries, dykes, hedgerows and intrusive trees. In addition, the weight of the engines caused a number of problems. When the ponderous machines lumbered from one field or location to another, they often left a trail of destruction behind them, sometimes knocking down gateposts, backing into walls or managing to terrify the livestock. However, these problems were not enough to prevent steam cultivation being widely adopted and although its onward progress was slowed when many farmers of heavier arable land went over to growing grass in a large way during the difficult years of the last quarter of the 19th century, steam remained an essential form of power on large numbers of farms until swiftly rendered obsolete by the internal combustion engine from the 1920s. The use of steam power on the land continued on a gradually reducing scale until the 1950s.

The next major mechanical development, and one that certainly turned out to be of prime importance, was the addition to existing reaping machines of a sheaf-binding function. A reaper designed by Patrick Bell had first appeared in 1828 but it was to be a generation before a machine fitted with a binding function started to come into general use. This was a rival machine designed by McCormick in the United States and was first successfully shown in this country at the Great Exhibition of 1851. Modifications were continuous and binders came into very general use once they were fitted with a string knotting device rather than one employing wire. Ironically the reaper-binder went on to have far-reaching effects on British farming because it proved eminently suitable for use on the immense prairie-lands of America where labour was in short supply. Huge quantities of grain, especially from the USA, were to flood into this country in the last quarter of the nineteenth century, undercutting the prices at which British farmers sold their grain and causing distress to many.

It would be naive to suppose that all farmers rushed to obtain the latest technology in spite of the generally favourable reception being accorded to devices which economised on labour costs. Many farms remained small and had neither the money to spend on mechanisation nor enough space to be able to derive any real advantage from the use of machinery. It was estimated that a land holding of at least 300 acres was needed for many of the new techniques to pay, yet in the 1850s, farms of this size constituted only around 30% of all cultivated land. The horse-drawn reaper, for example, was not used widely until the 1850s and even after that time the traditional scythe, sickle or hook remained commonplace for many years. Many farmers were markedly reluctant to change. Agricultural wages were low and this was a disincentive to major investment in machinery. Additionally, many farmers had very close and sentimental links with their workers and their horses and could not happily contemplate replacing them with machines.

The Peterborough Shows continue to expand

Agricultural shows like those held by the Peterborough Agricultural Society provided a very important shop window for the manufacturers of agricultural machinery and implements. At first, the exhibitors and competitors would tend to be the local blacksmiths and engineers whose operations were on a small scale and whose workshops could be found in and around almost any town or large village centred particularly in the areas of arable farming. As so often happened, take-overs and mergers were to result in the concentration of large amounts of the industry's capacity in the hands of a relatively small number of large companies which became virtually household names. Mention has already been made of how the Peterborough firm headed by W.P. Stanley was to evolve into the nationally acclaimed firm of Barford and Perkins.

It was during the Show of 1858 that the steam plough first made its appearance in the City. This successful Show also included the revival of the horse event after a lapse of several years. The President, the Hon. George Fitzwilliam reported that he had been to see the Northants County Agricultural Show at Northampton and that their Secretary had suggested that the two Societies' respective shows should amalgamate. The Peterborough Society members, when consulted, were unimpressed by this suggestion and voted against the proposal, it being stated rather ungraciously that the Peterborough Society would soon be larger than the Northants one anyway.

The 1860 Show was memorable for having taken place on what was recorded as probably the most dismal day of the most dismal year for a decade or more. Given the relentless and torrential rain which was a feature of that month, it is perhaps a miracle that any kind of show was staged at all. However we hear, for example, of a ploughing competition that took place in a large clover field belonging to a Mr

Charles Little near what was then called Spittle (sic) Bridge. This competition was described as being of an 'excellent standard'. The competition for other classes took place in a field belonging to Messrs Parker and Hopkinson near the five-sail windmill located in the Lincoln Road at Millfield. By a strange coincidence, this land turned out to be part of the Society's showground from 1887 to 1910.

The Show of 1861, unlike the previous year, enjoyed extremely fine weather and as a result there was a larger than usual attendance. Again the ploughing contest took place on Mr Little's land near Spittle Bridge but with so many entries, a field belonging to Mr R. Taverner near the present Westwood Bridge also had to be brought into use. Like many prominent Peterborough citizens of the period, Mr Taverner was later to have a road named after him. The main show of stock was held as usual, at the 'Waggon and Horses' Showground, but with the increase of entries and growing attendance, the need for a bigger showground was becoming ever more apparent. The quality of the livestock entries was said to equal any seen in past shows. Among the most notable were splendid animals entered for the fat oxen class by Earl Spencer, two of these taking major prizes. In the implement classes, Amies and Barford were the largest exhibitors and they carried off all the prizes offered. During the Society's customary Annual Dinner held after the show, there was a long debate about the growing need for a sufficiently large cattle market in the centre of Peterborough and the members resolved that the finding and securing of a site for such a market was a priority for both the City and the farming community of the area.

The Show of 1862 was heralded in the *Peterborough Advertiser* as the 25th Anniversary Show. This clearly emphasises the unanswered questions harboured in historians' minds about the continuity of the Annual Shows during the Society's early years. During this show, Amies and Barford were the only exhibitors in the implement classes and therefore carried off all the prizes for the second year running. No reason was given for such a scanty level of entries. We do not know whether it was the result of lack of interest or because potential competitors thought it was a foregone conclusion that Amies and Barford would win everything anyway, thereby rendering entry a pointless exercise. This year the press seem to have been somewhat disgruntled, complaining at length about the difficulty of getting into the Showground to compile their reports and they reiterated the urgent necessity for finding a larger and more suitable location for the staging of the Show.

At the Annual Dinner held after the show, the President, the 6th Earl Fitzwilliam, returned to the theme of the urgent need to find a site for the new cattle market and he made the point that the existing market which was held in the streets of the City Centre was

> *... so busy that on market day there is not even room for a man and a wheel barrow to travel up the main street.*

As a direct result of this overcrowding, a 'Peterborough Cattle Market Company' was formed and a site was eventually purchased from the 6th Earl Fitzwilliam

9. View of the Cattle Market, Peterborough.

This photograph taken about 1910 shows the Cattle Market which came into use in 1865 on the land purchased from the 6th Earl Fitzwilliam. Its location was close to the current retail market and the adjacent multi-storey car park. Its name is recalled at the present time by Cattle Market Road.

himself. The resulting market was opened in 1865 on land which is close to the current retail market and the erstwhile 'Embassy' Theatre in Broadway.

The Show of 1864 was considered a great success and a little light relief was provided in the livestock part of the show when a sow was mistakenly entered in the boar class and even more mistakenly won it! For the first time an associated poultry show was held at the Corn Exchange. This building had been completed in 1846 and was a prominent City Centre landmark standing next to St John's Church and close to the Market Place. The main show was held for the very last time at the 'Waggon and Horses' Showground. At their Annual Dinner after the Show, the members agreed that the Peterborough Agricultural Society should unite with the Northampton Agricultural Society in order to stage a combined Agricultural Show in Peterborough for the first time in 1865.

This was the end of another era in the development of the Peterborough Agricultural Society and a cause for celebration. The Society had established itself firmly with the farmers of the district and with the local community and its shows were so successful that they had outgrown the available space. It had been successful in its encouragement of innovation and improvement in the agricultural practices of the area and its Shows had also become a popular spectacle for the general public. People eagerly flocked to see machinery, particularly if it was actually working and steam-powered. Also such items as parades of horses were always popular with spectators and so gradually the idea of entertainment became a major consideration on the part of the organisers. The entrance fees of the general public came to provide the Society with some of the income it needed effectively to play its role as one of the country's leading Agricultural Societies.

Unquestionably the machinery and stock competitions and trials that were held at the Shows put on by Agricultural Societies, and in particular the larger and the more prestigious ones, helped to stimulate innovation in the industry as a whole. Those individual companies and farmers who participated were thereby given the necessary encouragement to think of implementing improvements in their working practices, with beneficial effects on agricultural productivity. For that reason, the contribution of Agricultural Societies as a whole to the development of British farming is quite inestimable and is a subject that has been somewhat underrated by historians. It is also true to say that agricultural societies in this country were neglected by government agencies. On the Continent, in Prussia and Denmark, for example, each province had a semi-official Society which acted as a co-ordinator for the various district and local Agricultural Societies. In several European countries, in the United States, in Canada and Australia, Agricultural Societies actually received substantial government funding by way of official recognition of their contribution to the growth of the efficiency and output of the farming industry.

In conclusion, the winning of awards at agricultural shows was clearly something that was seen as conferring considerable prestige on their recipients. Because of this, it was by no means unknown for deceit and sharp practice to occur. Amongst such skulduggery was the illicit feeding of competing animals with large amounts of linseed cake, and, at least so it was alleged, gin, cream and aniseed, and these must not only have assisted rapid weight gain but also induced a sense of euphoria and general bonhomie in the competing animals themselves. Sometimes pigs' teeth were filed in order to make the animals look younger. Alas we shall never know the full truth about all the nefarious and underhand activities that took place behind the scenes at the Peterborough and other Societies' Agricultural Shows in the nineteenth century. Would anybody have the temerity to suggest that similarly dastardly deeds take place at modern shows?

CHAPTER THREE

The Period of Mid-Victorian Innovation 1865 to 1880

The middle of the nineteenth century was a period of rapid change in British agriculture and local evidence of this was the decision of the Society in 1865 to build on its success and to move to a new and more extensive Showground at Boroughbury, located just north of the City Centre. The land concerned consisted of eight acres (twice the size of the previous Showground) and some of this was on the site of the old Boroughbury Manor House and the rest on land belonging to the Fitzwilliam family and to Mr Little, himself a member of the Society.

10. The Junction of Geneva Street and Lincoln Road.

A street scene dated 1907 showing urban development where, in the background, part of the Boroughbury Showground had previously been sited. Notice the tram belonging to the Peterborough Electric Traction Company passing 'Rothesay Villas' in the distrance. They had been built partly from material reclaimed when the medieval Boroughbury Barn, which stood in Church Walk, was demolished in the late 1880s.

The decision to undertake the move was an opportune one because separate discussions were already in progress between the Northamptonshire Agricultural Society and the Peterborough Society with a view to close collaboration or even a merger for one year for the purpose of staging the Northamptonshire County Agricultural Show in the City in 1865. This show moved round the county from venue to venue and a visit to Peterborough could reasonably be seen as overdue. The committee working on this project was considering the use of a suitably-sized plot of land near the Eastfield Road cemetery but were so impressed with the possibilities of the Boroughbury site that they readily accepted the invitation extended by the Hon. C.W. Fitzwilliam to hold the show there. The 1865 show, then, was a pioneer joint venture and took place on the new Boroughbury Showground on Wednesday and Thursday, 5 and 6 July.

The local committee considered the move to the new site and the collaboration with the Northamptonshire Society as being of particular importance to the City and, by way of celebration, decided to erect a large floral arch in the Market Place, paid for by public subscription. The *Peterborough Advertiser* of Saturday, 8 July describes the arrangements for the decorations in the following way:

> *The arch in the Market Place, extends from Mr Swallow's on the one side, to Mr Bristow's on the other. This arch was principally composed of evergreens, surmounted by a plough under which was the motto, 'Speed the Plough', painted in letters on a black background. The effect of this arch was not very lively, owing to the use of this background for the lettering. On the opposite side was a gilded crown, and from various portions of the arch gaily coloured flags were flying. Decidedly the best and most elaborate triumphal arch was that erected at the expense of Mr Royce, builder, in Bridge Street. It was neat and simple in design, was striking in appearance, and the flags, emblems and mottoes were interspersed with artistic effect. In the centre was the word, 'Welcome', on each side 'Speed the Plough', and 'Success to Agriculture', and suspended were two ploughs and other implements of husbandry to be found upon a farm. At dusk hour, the arch was illuminated by gas, which gave it a brilliant appearance. A third arch, rather a curiosity in its way, was erected at Fletton Gate, and formed a capital advertisement for its designer. It was constructed by Mr Sellers and was composed entirely of osiers. Amongst the private decoration of buildings, was the Stamford and Spalding Bank, which had two large flags hung from the roof of the building, and the motto, 'Prosperity to Agriculture', worked in gilt letters on a red background, evergreens and roses forming a border. This decoration was much admired. Other buildings were similarly decorated with mottoes such as 'Welcome to the City', 'A Bountiful Harvest - a Nation's Blessing' and 'Success to Agriculture.' Long Causeway, Church Street, Priestgate, Bridge Street, Narrow Street, Midgate and Westgate were similarly decorated, in fact all the streets leading to the Show Ground were covered with flags and bunting.*

On the evenings of Wednesday and Friday 5 and 7 July, firework displays had been held in Tout Hill Field near the Cathedral to support and advertise the County Agricultural Show. The fireworks were accompanied by a number of hot air balloon ascents. At that time the mysteries of lighter-than-air flight were being investigated by several daredevil aviation pioneers and their activities could always be guaranteed to attract sizeable crowds. Another associated event of interest was a cricket match in which a local team of young bloods took on 'The United All England Eleven'.

The Showground not only utilised the new Boroughbury site but made use of an

11. Decorative Agricultural Arch, Broad Street, Peterborough, 1865.

This triumphal arch, built across Broad Street, was paid for by Mr. Royce, a local shopkeeper, and is seen here in the process of construction. In the centre is the word 'Welcome' and ready to be lifted into place on each side are two ploughs with the slogans, 'Speed the Plough' and 'Success to Agriculture', respectively.

additional two acres of adjacent land belonging to Mr Little, himself a prominent member of the Peterborough Agricultural Society who has been mentioned before.

Peterborough's excellent railway links ensured that exhibitors and competitors from across the country and other participants had easy access to the Shows and 1865 was no exception. Even given the then high admission price of half-a-crown, the organisers were extremely pleased with the number of paying visitors, markedly up on previous years.

The display of the latest farming technology drew appreciative crowds. Mr C. Burrell, of Thetford, attracted a great deal of interest with a powerful and imposing traction engine which he had used to pull a 'thrashing' machine and straw elevator to Peterborough all the way from his works in Norfolk, a distance of fifty miles. This was recognised as a major feat by the standards of the time. One of Peterborough's own agricultural machinery-making companies, that of Amies and Barford had the largest display in this category of exhibits and Mr

Hayes of Elton again put on view his widely acclaimed straw elevator and grinding mill. The livestock show was generally considered the best ever presented at the Northamptonshire County Show and the *Advertiser* commented rather curiously: *Whether in size, symmetry [sic], or quality the animals could not be surpassed.*

The floral and horticultural parts of the show also proved a great success, the highlight being a magnificent Australian Tree Fern (*Dicksonia Antarctica*) from the Fitzwilliam Estate at nearby Milton. It weighed at least a ton, required six men to lift it and was considered to be the largest and most symmetrical plant of its kind in England. The ploughing contests and the awards to labourers were held on 7 October on some fields owned by Mr Henry Little and which were situated at the back of the Peterborough Gaol, built in 1842. Part of this building is now 'The Sessions House' pub and restaurant in Thorpe Road near the current District Hospital. During the Annual Show Dinner it was stated that:

> *The introduction of the steam engines and steam thrashing machine into the farmyards has certainly raised the intelligence of the farm labourer. A man could not be set to manage a steam engine without becoming more intelligent.*

This was an interesting comment in its context because it was not just the employers' side of the agricultural industry that was becoming aware that the development of steam power and the use of increasingly sophisticated machinery required more skill and training on the part of those who worked and serviced the new equipment. This awareness was filtering through to the workers as well and was to play a major part in the struggle that was soon to develop around the right of farm workers to organise in trade unions to negotiate collectively on crucial issues such as wages and conditions.

Agriculture was still in the phase conveniently if loosely described as 'High Farming' which it will be remembered was an intensive type of farming, involving investment in the latest techniques to maximise output and productivity. This policy worked well so long as markets kept expanding, prices were favourable and labour remained cheap. Indeed all these factors remained encouraging into the early 1870s. The urban part of the population, already over 50% in 1851, continued to grow and by 1881 actually numbered twice the rural population. At this time there was very little foreign competition, labour costs remained low, wheat prices were still fairly stable and the price of other commodities such as beef, lamb and cheese were rising quite rapidly in the early 1860s. Labour costs were still low enough in the south of England to act as a real disincentive to the introduction of farm machinery.

With the combined Peterborough and Northants Show of 1865 and the new Showground proving to be such a success, not only with exhibitors and competitors but the fee-paying public, the Society's officials and organisers of the Annual Show could be forgiven for thinking that a new era of continued growth and prosperity was beckoning them. However this was not to be, owing to the ravages of the highly contagious cattle plague or rinderpest which, having swept through Europe, now hit the British Isles, most cattle-breeding areas of the country being severely affected. An embargo was placed on the movement and exhibition of cattle and this undoubtedly affected the impact that the Show was able to make, albeit through no fault of the organisers. It was not until 1868 that cattle were again entered and judged at the Peterborough Show.

The Show took place in 1866 and 1867 but on a reduced scale owing to the absence of the bovine entries. At the same time the practice continued whereby certain of the events, particularly the ploughing competitions, were held away from the main Showground. For example, in the case of 1866 and 1867, the ploughing exhibitions were held at Mr Wagstaff's farm in the parish of what was then locally called Woodstone, just across the River Nene from the City.

Despite the setback resulting from the outbreak of rinderpest, the Society went ahead to build on the results of the particularly successful amalgamated Show of 1865, especially in terms of promotional and display material. Most spectacular were the series of arches erected at the main entrance in Lincoln Road. In the centre arch there was a plough and the motto, 'Unity is strength,' while the outside arches sported the apt inscriptions, 'Agricultural' and 'Horticultural', respectively. A number of improvements had been made to the Boroughbury Showground from which it is clear that the organisers were constantly looking to widen the appeal of the Show and the quality of the services provided so as to attract not only more agricultural exhibitors and competitors but larger numbers of the general public intent on having a good day out. Their success is shown by the way in which the Peterborough Society, although strictly speaking only a local Agricultural Society, was already developing the infrastructure and reputation of a society of county or even regional status. There was an increasing range of exhibits which had little or nothing to do with agriculture and while some of the purists among the Society's members felt that the Show was moving away from its stated purposes, it is clear that the fee-paying public wanted to be entertained and stimulated by a diversity of attractions.

In 1869, the *Peterborough Advertiser* evidently approved of efforts being made to attract more ladies to the Show:

> *Among the attractions of the ground were several 'Boudoirs', as we may call them, so nattily were the stands fitted up in which Messrs Inslip and Stanford showed the Britannia Sewing Machine, Mr Clarabut, Wheeler and Wilson's sewing machine, and a magnificent collection of silver cups, ranging from 3 shillings to 20 guineas each. He also showed what is certainly a novel thing at an Agricultural Exhibition, a case of diamond and other precious stone ornaments. Travelling bags, dressing cases and fancy goods were also there in abundance. A novelty wine cooler, 'The Pallash', was also on sale.*

While the general tone of the *Advertiser's* comments is one of approval for the efforts of the Society, clearly the local paper had never thought much of the ornamental arch at the entrance to the Showground and in noting its removal, adds somewhat tersely, the words: *'wisely, as we think'*. Shortly after what was agreed to be a very successful Show, the Society had its annual Show Dinner and in a large marquee, 250 guests sat down to the meal. In his after-dinner speech, The

12. Sale Notice, 1868.

A notice relating to the sale of surplus materials from the Boroughbury Showground.

President, The Hon. G.W. Fitzwilliam, reminded the Society members of past Shows and he spoke specifically about the 1861 Show that had been held at the old 'Waggon and Horses' Showground. His theme was the great improvements that had taken place since the move to Boroughbury but we can discern a certain wistful and nostalgic affection for the old days:

> *All would remember where they held their meeting in 1861. It was not on this Showground but in a little poky yard, hardly large enough to swing a cat. There was no room to show the merits of either horse or beast, and scarcely room to do as much for a pig. They had an audience that did not show their appreciation in the most polite manner. It was generally composed of boys, who dangled their dirty little legs down the side of the wall, and kept up a chorus with the heels of their boots, jibing and jeering everyone who did not show an animal to what they thought was an advantage.*

In 1871 the Peterborough and the Northamptonshire Agricultural Societies once again agreed to stage a joint Show in the city. The very high regard in which the Hon. G. W. Fitzwilliam was held was shown when he was singled out for the unprecedented honour of election as President of both Societies.

The previous joint show in 1865 had been held at Boroughbury but on this occasion because of the earlier Show date, it was felt that the site would not stand the wear and tear occasioned particularly by the working machines and implements which would be exhibited. As a consequence it was decided to house the horticultural part of the Show on Tout Hill Close Field, just north of the Cathedral. The main agricultural part of the Show was located on twelve acres of land bordering what was then called Thorney Road and which is now Eastfield Road near its junction with Padholme Road and adjoining and embracing the old Cemetery.

Unusually, this particular Show was held in September but the change of venue and time of the year does not seem to have harmfully affected either the competitors, exhibitors or paying visitors and indeed there were comments that the livestock exhibited was the best ever seen. Among the outstanding exhibits was a splendid bull belonging to the Marquess of Exeter. This bull was a formidable creature and was already something of a celebrity in his own right, proudly possessing the stirring name of 'Telemachus', the great adventurer who, in Greek mythology, was the only son of Odysseus and Penelope. The equine competitors were generally reckoned to be outstanding and perhaps the greatest impact was made by a superb five-year old gelding belonging to Mr William Whitehead from Woollaston. This was called 'Little John' and was being exhibited for the very first time. As with many previous Shows, the implement part of the proceedings was dominated by the local firm of Amies and Barford whose exhibits covered one complete acre of this rather makeshift Showground.

Despite the successful Show at the Thorney Road site, in 1872 the Society reverted to land belonging to Mr Little at Boroughbury. This was not on the original site of the old Boroughbury Showground because some of this land was now required for building purposes. The growing prosperity of Peterborough was increasingly leading to business and professional people looking for desirable residences nearby and so new housing was beginning to encroach onto what had hitherto been a largely pastoral area north of the City Centre.

The *Peterborough Standard* described this latest site in favourable terms:

> *The Agricultural Showground in the largest meadow at Boroughbury is an extensive one forming nearly a square, and is boarded on three of its sides, and has extra entrances north and south. The gravel road from the Lincoln Road to Allan's Lane divided the Showground into two portions. The west portion is entered from the Lincoln Road and is assigned to the horses, the boxes for which form the west and north line of the enclosure, leaving an ample piece of green required for the judging and jumping area. The boxes built were of the most approved style, and were covered. On the east side of the ground a spacious covered gallery for spectators was erected. Tents, pens and stalls were also assembled to support the various needs of the show animals and smaller exhibits.*

In 1873 this modified Showground site at Boroughbury was stated as covering between 7 and 8 acres but by 1874 this had increased to about 9 acres and a number of improved facilities had been added. It included a spacious new ring, a dedicated building for the poultry part of the Show and a main grandstand with accommodation for 300 people. The ring used for the horse trials was about 200 yards by 60 yards and contained three constructed jumps. The first of these was made of hurdles covered in gorse; the second and principle one, was more formidable, having a movable bar adjustable to a maximum height of 5 feet. The third was a double jump also consisting of hurdles and gorse.

The Annual Dinner of the Society in 1874 was notable because the guest of honour was Henry Pearson Gates, the City's first Mayor. Peterborough had been granted the status of a Borough by the Queen earlier in the year and this dinner was Gates' inaugural public engagement as Mayor.

In May 1875 the Royal Agricultural Society of England held its regular meeting to decide on the location of its next Annual Show. This Show, the biggest of its kind, was peripatetic and the Society tried to ensure that all parts of the country were visited on a reasonably equitable basis. The two venues under discussion on this particular occasion were Peterborough and Birmingham. The Peterborough delegates were a strong team composed of Mr William Barford, The Marquis of Huntly, the Mayor (the aforementioned H.P. Gates) and The Hon George W. Fitzwilliam. The voting was 27 to 12 in favour of Birmingham. This was a blow to the Society and the City, both of whom could have expected a major influx of visitors with their attendant spending power. However, the Royal Society's members knew by experience that they attracted the most visitors when they located the Show in one of the larger urban venues and they needed the extra revenue that was likely to be raised by holding the event at Birmingham with its greater population and more central location.

A feature of the 1875 Peterborough Show was an exhibition of Messrs Barford and Perkins improved steam cultivating machinery and this was held on a field close to the Boroughbury Showground. On the main ground itself there was a special feature given over to vendors of cattle foods and medicaments. These included the quaintly-named 'Ground Hope's Cattle Food' and a patent nostrum and guaranteed cure-all for cattle and other animals with an even stranger name, 'Tipper's Medicated Mystery and Arabian Cattle Cake.'

In 1876 and 1877 the continued encroachment of new housing and commercial property around the Boroughbury area was threatening the Society's continued use of the Showground not least because the rents for the land in this area were becoming increasingly burdensome. A warning of possible problems was clearly given at the Society's annual meeting on the 3 January, 1877 according to the *Peterborough Advertiser*:

> *Although your committee were not able to secure the ground upon the same favourable terms as in former years, still it is hoped they will be considered to have exercised a wide discretion in securing a site so close to the town, and well adapted to the purpose, and where so many successful shows have been held.*

It was to be just five years before the Society finally and reluctantly had to accept that its Showground was being overtaken by Peterborough's urban expansion. It then needed to find a location that was on what we would now call a 'greenfield' site with more space for expansion and not threatened at the time by alternative use for building purposes. A precedent had been established with the very first move from the 'Waggon and Horses Yard' and the same process was repeated with the subsequent move to the Boroughbury sites and later moves to Millfield, Eastfield and eventually Alwalton. In each case the Society's success meant it required more land for the Show but the rapid growth of the City ensured that the necessary space was simply no longer available. It could be that at some stage in the future we may see the process repeated with regard to the current Showground at Alwalton where most of the land is freehold.

In 1878 the 6th Earl Fitzwilliam played a key role in the formation of the Peterborough and Midland Counties Foxhound Society. Almost immediately it decided to collaborate very closely with the Peterborough Agricultural Society and to stage its own Show annually on the second day of the Agricultural Society's Show and at the same venue.

The close relationship between the two Societies was to prove of great mutual benefit encouraging many prestigious visitors, including members of the Royal Family, to attend, thereby widening the attraction of the Show in general and encouraging more people to visit. The Prince of Wales, later to be King Edward VII had agreed to be the Foxhound Society's first patron and he attended the opening event. The *Peterborough Advertiser* recorded the importance of the close collaboration of the two societies and the likely effect on attendances in the following way:

> *The Peterborough Agricultural Society held its Annual Exhibition in the Boroughbury field on Wednesday and Thursday last. On the second day the Peterborough and Midland Counties Hound Show took place, thus giving an importance to the occasion which it was impossible to overestimate. In connection with the latter the visit of the Prince and Princess of Wales to the City increased the interest tenfold, and consequently nothing was wanting to surround the proceedings with attractions of the highest order.*

The visit of the Royal couple certainly caught the imagination of both the general public and tradesmen and the City streets and public places disappeared under an absolute welter of bunting, decorations and slogans. One particular example worth recording was that executed by a Mr Hortor, a local perfumer. He erected a large banner adorned with the following rather sycophantic lines:

> *May he! Our prince continual rise,*
> *And bear his honours to the skies,*
> *While dazzling out from History's line,*
> *Our loved Princess's virtues shine.*

What Their Royal Highnesses thought of this doggerel was not recorded, which may be just as well.

The 1878 Show was widely considered to be one of the most successful yet but if anything it was eclipsed by that of 1879. Once again the Prince of Wales visited the Show on its second day and the Peterborough Advertiser recorded his arrival in the City at the Great Eastern Railway Company's station just over the Town Bridge on the south bank of the River Nene:

Shortly before two o'clock the cry was given, 'Here he comes', and the special train slowly steamed into the station. His Royal Highness at once took a seat in a horse drawn carriage beside the Marquis of Huntly, and was driven through the City. The streets were crowded and the Market Place full of people. The carriage then passed rapidly down Long Causeway and on to the Fitzwilliam Hall near the rink in Park Road and so into the Showground itself.

The Fitzwilliam Hall, built in 1872 was where the current Central Library now stands in Broadway and had been specially fitted out for the occasion while a portion of the adjacent Theatre Royal had been transformed into a handsome reception room for the Royal party. Among the innovations at this Show was a new stand appropriately entitled 'The Prince of Wales Stand' while a telegraphic device was brought into use for the notification of the winners of the various competitions.

In 1880 the Northampton County Agricultural Society agreed to join with the Peterborough Society to stage another joint Show in the City but overtures to the Huntingdon Agricultural Society to add their weight to an amalgamated Show were politely but firmly rejected. The joint effort created a Show that was obviously considerably larger than normal and by having done so, it provided just the necessary impetus for discussion of a move to another bigger venue to become a part of the Society's agenda.

The national context of Agriculture in this period

The good times for agriculture were showing signs of coming to an end in the middle of the 1870s. The market for farm produce was continuing to grow rapidly as were real wages, at least among much of the urban population. A consequence of this was that there was a substantial rise in the consumption of meat and dairy products but less was being spent on bread. The really unfortunate fact from the point of view of many British farmers was that while there was an expanding market for foodstuffs, it seemed that an inexorably growing share of that market was being captured by imports from overseas. Grain from Russia and the Americas was pouring in and bringing the price of wheat down to the lowest levels for 150 years or more. Additionally, large quantities of Danish butter and bacon were reaching the tables of middle class consumers while frozen beef and mutton from the Americas and Australia found a ready market among working class consumers. Increasing amounts of fruit and vegetables were finding their way into this country from Europe. This was, of course, a welcome development for the British consumer who was now enjoying an unprecedented range of cheap food products but gloomy news for much of the country's farming community.

Such is the background to the so-called 'Great Depression' in agriculture which can be seen to have been a delayed reaction to the repeal of the Corn Laws in 1846 and the commitment to the concept of free trade. This was to become a canon of British economic policy, although an increasingly controversial one, and was the product of the country's early lead in the process of industrialisation which meant that for a period of some decades in the middle of the nineteenth century, Britain was the undisputed world leader in economic, industrial, commercial and political power.

In 1851, when Britain's economic fortunes might rightly be said to have been at their peak, agriculture employed more than 20% of the working population and produced approximately the same percentage of the national income. However, 50 years later it employed less than 10% of the labour force and its share of the national income had fallen much further, to less than 15%. In this situation the industry was forced to undergo the painful process of adjusting to harsh new realities. Numerous long-established estates were auctioned off and then broken up. Much of the poorest arable land was abandoned and the workforce was cut back. Many of the youngest and potentially most able farm-workers either left the land for jobs in the towns, or emigrated.

The picture was not one of total gloom. Certainly the grain farmers of the eastern and midland counties were hard hit when their prices and profits fell sharply and remained low but elsewhere livestock producers, dairy farmers, market gardeners, potato growers and fruit farmers were faced with a buoyant, expanding market for their produce.

It was also not helpful that the weather in the last quarter of the nineteenth century proved to be so capricious. The years 1875 to 1878 were characterised by extremely wet, cold summers with consequent poor harvests and shortages of winter fodder. The excessive damp led to outbreaks of pleuro-pneumonia in cattle and liver-rot in sheep during the years 1879 to 1882. Highly damaging droughts and severe frosts followed in the years 1885 to 1887. Later, in 1891, there was another very wet summer followed by further very severe droughts. Then the winter of 1894 proved to be one of the coldest in memory.

In the early 1870s, there had been a short-lived but acrimonious attempt to develop trade unionism among the country's farm workers. Trade unionism was growing nationally among the male workers in a variety of skilled trades such as the engineers, printers and carpenters. Although success in trade disputes was certainly not one-sided, the creation of these unions was a clear demonstration of the potentially increased strength that came from combining for the purposes of collective bargaining. At the same time, the unions achieved some legal recognition in the eyes of the courts.

The 'High Farming' of the mid-1840s to the mid-1870s had been a real boom period for those farmers who were best able to take advantage of the opportunities that were presented at that time. Socially the period was a divisive one because some of those who became affluent went to considerable lengths to distance themselves from their employees. A three-tiered class structure began to emerge in the countryside; of landowners, tenant farmers and land-less farm labourers and while for those on the upper and middle tiers of the hierarchy, this may have been seen as the natural and desirable order of things where each and everyone knew his place, the more intelligent and better educated labourers argued that a greater proportion of the fruits of this new prosperity should filter down to them.

In the early 1870s, a survey of land-ownership in the British Isles was carried out, known as the 'New Domesday Survey'. It revealed, perhaps not surprisingly, that no more than 7,000 men owned 80% of the country's land and, as a consequence, they enjoyed a position of absolute domination not only in rural society but in many cases in the wider social, economic and political hierarchy. This survey defined an aristocrat as the owner of an estate of over 10,000 acres, which was described as a 'great estate', and estates of this sort occupied one quarter of the land surface of England. It is interesting to note that by this definition, nearby Rutland was the most aristocratic of counties with 53% of its land occupied by great estates.

Below the aristocracy in the hierarchy of land-ownership, were the gentry, numbering over 3,000 and in possession of estates of between 1,000 and 10,000 acres. The gentry assisted the aristocracy to administer justice, providing the majority of the magistrates and the officers for the county's militia and, at least after 1834, they had a major say in the localised workings of the Poor Law. While generalisations in history are always risky, the gentry's ownership of land and their social aspirations tended to place them very firmly with the aristocracy. However some of them had a 'hands-on' approach to the everyday problems that occurred on their estates and they would frequently rub shoulders with lesser tenant farmers and in some cases even with the labourers themselves.

The 'New Domesday Survey' identified 'yeomen' as those working holdings of between 100 and 1,000 acres and they, even more than the gentry, found it hard completely to disassociate themselves from the interests of those they had to work closely with but whose social status was much lower.

It is appropriate to make a distinction between the predominantly pastoral north and west of England and the mainly arable south and east because the nature and conditions of village life varied widely between them. In the south and east, great use was made of hired labour and this made for a decisive social gulf between farmers and farm workers. In the largely pastoral areas, however, where much less use was made of hired labour, the most obvious social distinction was between landowners and tenants. In much of the East Midlands and East Anglia therefore, the rural community was overshadowed by the divide between the hired hands who largely lived in the village and the farmers and especially the landowners who increasingly tended to live away from their landholdings. In both a geographical and a social respect, therefore, a sense of community was gradually disappearing.

Intensive arable farming required a large workforce but the farm labourer found that, despite being required to use more complex skills and to assume greater responsibilities, he still occupied a very lowly position in society's hierarchy and that the relationship he had with the local aristocracy and gentry was definitely a subservient one. Understandably this did much to engender an attitude of 'us and them', a rigid observation of the divide between employer and employee.

The power of 'them' was evident at every turn in rural England. The farmers and landowners had a near-monopoly of employment opportunities but were also to be found having a major or controlling interest in such important institutions of rural life as housing, schooling, the Church, poor relief and the dispensing of justice. All this was a far cry from the roseate picture of social harmony that can be found in some of the paintings by contemporary artists such as Charles Richard Havell and William Maw Egley in his *Hullo, Largess! A Harvest Scene in Norfolk* (1860).

A long-running grievance of rural labourers and their families was the poor state of their housing which stood in the starkest contrast to that of the landowners and substantial farmers. Between 1851 and 1861 over 5,000 rural cottages were inspected. Extensive evidence was found of serious overcrowding which led to fears of the moral consequences, but there were also leaking roofs, dampness, infestation and various other environmental and health hazards. Great concern was expressed by well-meaning liberals of the degradation that this caused while others voiced the fear that, unless something was done, the rural poor might move into action over their grievances. Unrest among the rural population was no novelty and was an indication of a social reality very different from the misleading image sometimes put across about rural deference to the squire and the parson. It was manifest not only in the so-called 'Last Labourers Revolt' of protest in the Fens in the early 1830s which involved violent and illegal direct action but also throughout the eighteenth and much of the nineteenth century in the very widespread and open support given to criminal activities such as smuggling and poaching which often had in them a large element of social protest.

Another significant factor in rural society was the influence of Nonconformist Christianity, most especially, Methodism and Primitive Methodism. These movements captured the imagination of those large numbers of rural workers and their families who sought in their hopes for a blessed existence after death some compensation for the disappointments and travails they experienced in their earthly life. Nonconformity was a powerful educative force which used both day and Sunday schools not only for the purposes of devotion but, by emphasising the coming together of minds, assisted its adherents to become more articulate, confident and socially aware. Although expounded by people who were usually of unimpeachable respectability, the body of ideas that were developed contained an underlying social critique leading to a telling demand for greater justice. Not the least part of its significance was the way it contrasted its set of universal upright moral principles with the secular, cynical and opportunist principles on which power in civil society was based. Inevitably it called into question the moral authority of the squire and the parson and was therefore highly critical of the Established Church. There had been isolated attempts to establish unions among agricultural workers in the 1860s, especially in Lincolnshire and Leicestershire, but considerably greater success was achieved in 1872 when the figure of Joseph Arch (1826-1919) burst onto the scene. The traditional isolation and notoriously low levels of education of the rural worker were showing signs of breaking down by this time, particularly with the effects of the railways, the coming of the Penny Post and the appearance of cheaper newspapers, all of these assisting the more rapid and thorough dissemination of ideas. While the proportion of rural as opposed to urban workers was falling, the rise of technology demanded the possession of a greater range of skills by those who still worked the land. However they continued to experience bad housing, frequent overcrowding, a poor diet, low status and low wages.

Into this difficult situation stepped the fervent Arch, a Methodist lay-preacher of mercurial temperament. He unusually combined great skill in the rural crafts of hedging and mowing with quite spellbinding oratorical ability and considerable organisational talent. His first success was to set up a union in his home county of Warwickshire and in 1873 his tireless efforts led to the bringing together of various disparate localised organisations into a united body, the National Agricultural Labourers' Union (NALU), at its peak boasting 150,000 members and having its main strength in the counties south and east of a line from the Humber to the Severn.

Branches of the NALU sprang up, making demands for higher wages and a shorter working week and accompanying these with threats of strike action which often won them major short-term concessions. Arch was an astute user of the communications media of the time and was able to win, at least for a period, the support of a number of influential journalists and newspaper editors. This gained the agricultural workers substantial public sympathy and assistance from a number of urban trade unions and radical MPs. The union was led by serious, sober, self-taught men, many of them Nonconformists and the meetings were laced liberally with prayers and hymns. The NALU emphasised the value of education, of sobriety and thrift and it espoused many what were then radical demands such as further extensions to the male franchise and the need to nationalise the land.

The employers' side of the industry was split between some who were openly hostile to the labourers' aspirations and those with some sympathy. All the employers were quite frankly amazed by the evident ability of men who had been widely regarded as bucolic clodpolls and bumpkins, so successfully to articulate their needs and to organise towards achieving them.

It did not prove possible to maintain the momentum of organisation for long and the Union's advance slowed up after some particularly acrimonious defeats. The employers struck back, most forcibly in East Anglia, as they entered a period of economic depression, employing blackleg labour and locking strikers out where necessary and going on to make an example of local union leaders. It was always going to be hard to organise workers of this sort who were relatively isolated and could not afford the higher membership subscriptions of the skilled industrial workers which gave them some resources to help them in hard times. This episode left a considerable feeling of bitterness in the rural community particularly of the midlands, the south and the east and the response of many of those who could do so was to leave agriculture to seek work in the towns, or to emigrate.

The success of Agricultural Shows was markedly influenced by the same factors that made for success or otherwise in the industry itself. Thus disease, weather, costs, prices and government policy all had their effects but it is very much to the credit of the officers of the Peterborough Agricultural Society that during the difficult years from the 1870s, when the arable farmers of the region were experiencing life in the doldrums, the Show continued to grow in size and reputation and indeed was about to embark on an exciting phase of growth and expansion on a new site.

CHAPTER FOUR

The Start of The Great Shows 1881-1886

The Northamptonshire and Peterborough Societies combined to put on another Show in the City in 1881 and this was held on a new ground, known as 'The Rink', which covered about 16 acres. This site embraced an area including Park Road, Fitzwilliam Street, Crawthorne Road, Brook Street, Cattle Market Road and what is now Burghley Square, all close to the northern edge of the City Centre. Incidentally, Broadway, which was eventually to run through the middle of the site used by the Rink Showground, only started to be developed in 1880 and it was some few years before its use and extension threatened the Society's activities in this area of the burgeoning city.

The presence of the Annual Show was always good for trade and the City's shopkeepers went out of their way to attract and entice the many visitors who strolled around the City Centre before or after visiting the Show. One of these traders was a Mrs Amies who had an ironmongers shop in Long Causeway and in 1881 her display so impressed the *Peterborough Advertiser* that in an editorial article they made this comment:

> *The artistic arrangements of her exhibits...combined with the ornamental character of the display, rendered it one of the most attractive features of the exhibition.*

From this it is clear that the local press regarded the displays put on by the shopkeepers as a vital part of the City's effort to market itself effectively for the thousands of potential consumers who flocked into Peterborough on these exceptionally important two days in its annual calendar. The same article went on to provide vital information for any historian researching the vexed question of the dates and exact locations of the Society's various Showgrounds in the City. It makes it clear that what was in effect a City Centre entrance to the Showground, was located in what was later to become Broadway, one of the City's major streets.

The outline plan of the Rink Showground and a description are given in the same paper:

> *It is only necessary to view the Showground on the Park Road to be convinced of the great improvement which the amalgamation of the two Societies has brought about. Whereas in former years seven acres of land had been found sufficient for the Peterborough Society's exhibition, on this occasion more than double that quantity, in fact 16 acres - had to be called into requisition. The Rink Field as it is called, which has generally been used, was added to by the improvisation of the Peterborough Cricket Club Ground (Crawthorne Road) and the recreation-ground*

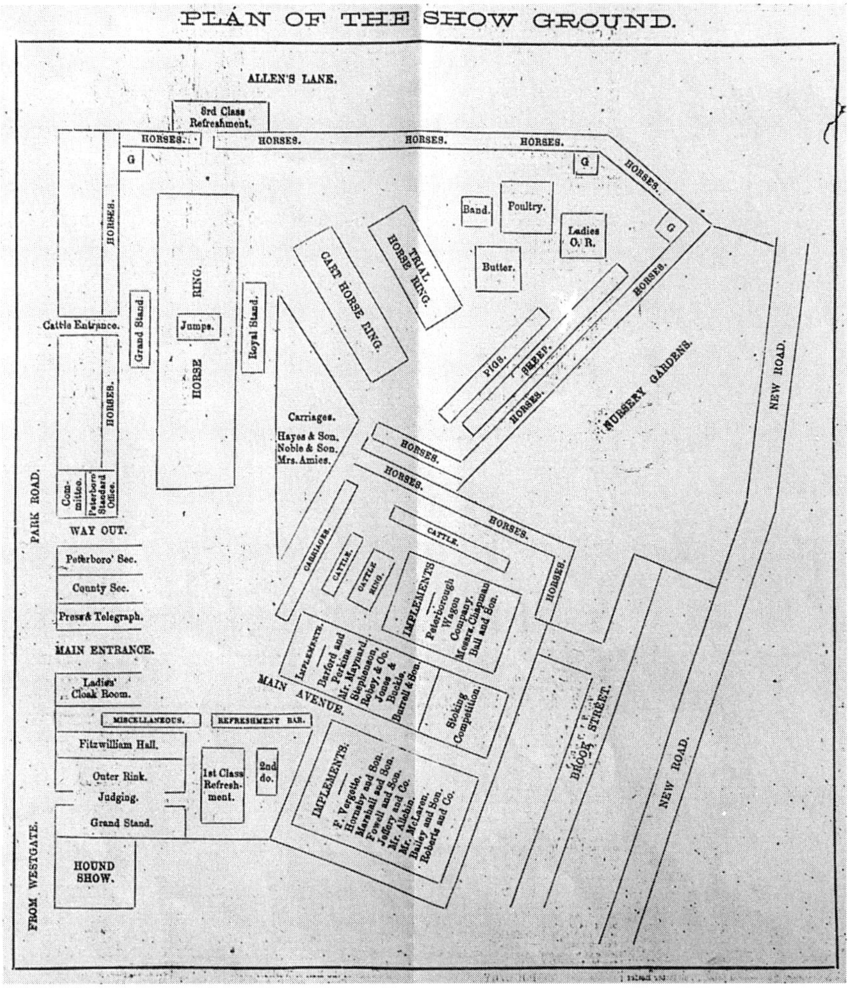

13. Plan of the 1881 Showground.

A plan of the Showground as used for the joint Show with the Northamptonshire Agricultural Society. This larger than normal Show display area embraced the 'Rink' Showground and various other parcels of land close by.

(Stanley Recreation Ground). Fences having been removed here and there in order to give free access to all parts of the show, it looked as a whole like an imposing and continuous collection of agricultural exhibits...The main entrance was in Park Road close by the skating rink and Hound Show marquee. At this point three separate gates were provided. Immediately inside the field to the right were the ladies' cloakrooms, whilst on the left were the press and telegraph and secretary's offices. Then what was described as the main avenue ran through the implements to the end of the recreation ground, where, owing to its distance from the stock, the machinery in motion was erected and steaming tests were conducted. A public luncheon tent, capable of seating 800 people, and various marquees, water jumps, stalls and a permanent grandstand and

Royal stand are also to be seen, with a main ring and an extra ring. The cricket ground pavilion was also used and sheep and pig pens were placed close to it.

The administrative and social centre of this new Showground was the Fitzwilliam Hall in Park Road. This had been built in 1872 by W.D. Nichols and was part theatre and part roller skating rink. It proved to be an effective reception area when pressed into service during the Prince of Wales' visit to the 1879 Show.

This Show broke all existing records in Peterborough in terms of the area needed to house the exhibits and also the sheer number of exhibits, some of which had to be displayed in the Market Place and the streets of the City. Certainly Barford and Perkins, the well-known local agricultural implement makers, exhibited not only at The Rink, but temporarily took over some land at Woodstone, belonging to a Mr Rowe. There they displayed working examples of such devices as grinding mills for grain, straw, hay and corn elevators and an 8 h.p. portable steam engine.

Back at the main Showground, there was much to interest all concerned. The equine entries were of an extremely high calibre and caused the judges in some cases to deliberate for over an hour before they could come to a final decision. Some well known animals such as 'Esperance', 'Evangeline', 'Rosamond' and 'Hercules' were on view. As far as the agricultural implements were concerned, all eyes were drawn to Walter Wood's harvester and self-binder. This had the great merit that, after the corn had been cut, it collected the grain and bound the straw into sheaves with string rather than the wire used by existing binding machines. The Royal Agricultural Society had already offered a gold medal to the designer of an effective string binding apparatus.

In many ways, the 1879 Show was a major watershed in the history of the Peterborough Agricultural Society because even before the additional presence of the Northamptonshire Society, a precedent had been struck for shows of larger and more ambitious format than those previously staged and which confirmed them as events that were of considerably more than local importance.

In 1882 the Peterborough Show, apparently for the first time, employed a military band to provide musical entertainment on appropriate occasions during the proceedings. This was the band of the 4th Derbyshire Regiment Sherwood Foresters. Their renderings were so popular that military music was to become a regular feature at subsequent Shows. On the second day, the proceedings were enlivened by an exciting display of fireworks by Mr James Pain, described in the local newspaper as *the celebrated London pyrotechnist.*

The growth and evident success of the Society's Annual Show and the prestige as well as the income that it brought to the City and surrounding district, were recognised when the Mayor of Peterborough, John Thompson, JP, issued a notice which was posted all over the City, requesting business places to close at 1pm on the Wednesday in order to give shop assistants, workpeople and others an opportunity to visit what he described as 'The Exhibition'.

The Society's officers were clearly not the sort of people who were content to rest on their laurels and the 1883 Show saw further improvements to the Rink

Showground. More sheds were added and a total of 350 horse boxes made available, extending around the whole of the rink field and even intruding into the adjacent Crawthorne Road cricket ground. A specialised structure was erected to house the engine-driving competitions and the accommodation in the grandstands was greatly increased.

Once again the local newspaper drew on the superlatives when reviewing the quantity and quality of the exhibits and described the display of pigs in the following manner:

> *Of pigs there has seldom been such a show at Peterborough. For size and breed they could not be beaten anywhere. Some were monsters of obesity, especially those fed with Mr Stephen Willson's Canadian Pig Powders...*

Mr Willson had a chemist's business in Long Causeway and must have revelled in this free publicity. In those days, at least so far as pigs were concerned, it was definitely size that counted.

In 1884 some modifications were made to the Rink Showground. It was impossible now to use the Cricket Ground which bordered Crawthorne Road because it had been fenced off securely. On this occasion the Society was able to obtain a small plot of land from Mr T.L. Barrett who owned the well-known Barratt's drapery shop which stood at the junction of Long Causeway and Midgate until comparatively recent times. Overall, about seven fewer acres were available for Show purposes in this particular year.

The show-jumping area was reconstructed - it had nine jumps as follows:

> The naked hurdle; A gorsed hurdle; The wide jump consisting of two sets of gorsed hurdles, with a pole between; A double jump; The stone wall jump; Double post and rails; The swing gate; Water jump, with fifteen feet of water; Adjustable height jump.

The champion in the equine events in 1884 was 'Safeguard', ridden by Master Tom Watson and which managed to complete the rounds with no faults.

The Society always held its Annual Stallion Show earlier than the main Show, in March or April. The 1885 Show was held on Saturday 21 March on part of the old Boroughbury Showground adjoining the skating rink and there were 66 entries. The entry classes were:

> 1. The owner of the best thoroughbred Stallion of any age, suitable for getting hunters.
> 2. The owner of the best Roadster Stallion, of any age.
> 3. The owner of the best Shirebred Cart Stallion four years old and upwards.
> 4. The owner of the best Shirebred Cart Stallion foaled in 1882.
> 5. The owner of the best Shirebred Cart Stallion foaled in 1883.
> 6. Champion Prize. The owner of the best Cart Stallion, exhibited in any of the classes.
> 7. Special Prize. The owner of the best Shire Cart Stallion exhibited in classes 3 or 4 which shall serve mares belonging to members of the Society.

The winner of this Special prize and the demanding if congenial duties that went with it was a splendid beast called 'Endymion', owned by Mr J.C. Allen of

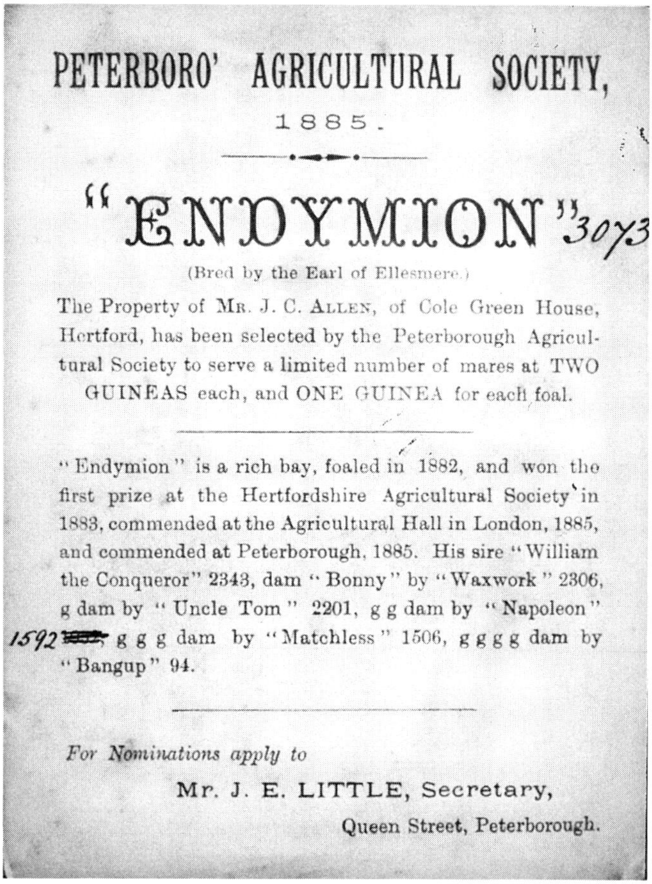

PETERBORO' AGRICULTURAL SOCIETY,

1885.

"ENDYMION" 3073

(Bred by the Earl of Ellesmere.)

The Property of MR. J. C. ALLEN, of Cole Green House, Hertford, has been selected by the Peterborough Agricultural Society to serve a limited number of mares at TWO GUINEAS each, and ONE GUINEA for each foal.

"Endymion" is a rich bay, foaled in 1882, and won the first prize at the Hertfordshire Agricultural Society in 1883, commended at the Agricultural Hall in London, 1885, and commended at Peterborough, 1885. His sire "William the Conqueror" 2343, dam "Bonny" by "Waxwork" 2306, g dam by "Uncle Tom" 2201, g g dam by "Napoleon" 1592, g g g dam by "Matchless" 1506, g g g g dam by "Bangup" 94.

For Nominations apply to

Mr. J. E. LITTLE, Secretary,

Queen Street, Peterborough.

14. Endymion.

A stud card dated 1885 advertising the services of a stallion called 'Endymion'. The Society acted as a facilitator for the services of this animal.

Hereford, and members could arrange stud agreements with the owner through the agency of the Society.

The Annual Show continued to go from strength to strength and in 1885 once again broke all previous records both for entries and paying entrants. The *Peterborough Advertiser* warmed enthusiastically to its theme:

> *Nothing succeeds like success...an axiom especially applicable to the Peterborough Agricultural Society's Show. Twenty years ago the Show was held in the 'Waggon and Horses Yard' in Bridge Street, and was, as may be imagined, of very insignificant character. Gradually, however, its importance has increased, owing to excellent management. This year there was an enormous increase in entries, the horses alone numbering 484 whilst all other entries exceeded any previous year, even the joint exhibition with Northamptonshire in 1881.*

In 1886 the 'Rink Showground' staged its final Annual Show. By this time the Peterborough Show had unquestionably become one of the country's premier Agricultural Shows and this fact was a matter of pride for the citizens of the City and its outlying districts as was made very clear by the *Peterborough Advertiser* in its editorials. In this year the *Advertiser* recorded:

> *The bells of the Parish Church rang merrily at intervals during the day, and many of the shop keepers closed their shops to give assistants an opportunity of visiting the Show... at the workhouse, children were allowed to visit the Show.*

At this Show considerable interest was generated by the working of a telephone installed by the South of England Telephone Company and by the Postal and Telegraph office from which numerous visitors sent sixpenny telegrams to friends and relations. Although this was not the first time that either telephones or telegrams had been seen in the City, their use was indicative of the manner in which the Annual Show was increasingly being seen as an effective 'shop window' for technological innovations and advances and as a place for the marketing of new products. Many factors affected this perception: the increasing numbers of visitors each year; their potential spending power and in some instances, their prestige and influence.

The last Show at the 'Rink' and the move away from sites close to the City Centre were evidence both of the steadily growing economic and commercial importance of Peterborough and the continuous expansion and spirit of enterprise and innovation which were marks of the Society's activities.

Agriculture faces a host of problems

The Society and its Shows continued to grow despite the hard times being experienced by many farmers in the last quarter of the nineteenth century. The prosperity experienced across much of the farming community in the middle of the century had been based on high prices for both livestock rearing and cereal production. The fact that Britain was a high cost agricultural producer had not been of great significance when little or no foreign competition was being experienced. By the 1880s, the industry found itself staggering under the impact of competition from low-cost producers in the Americas, Australasia and Eastern Europe. Even in Western Europe, countries such as Denmark and the Netherlands began to specialise in the production for export of high-value food products including butter and bacon.

Other countries in Western Europe preferred to protect their agricultural industries by imposing tariffs on imports, especially in periods of depression. The economic policy of successive British governments was that it was in the country's interest to maintain free trade. It was believed that cheap food imports indirectly kept down the cost of industrial workers' wages. That such a policy might have harmful effects on the agricultural sector was thought to be a small price to pay. The long-term continuance of this policy was a very clear indication of the changing balance of forces in British politics and the affirmation of growing power in the hands of the manufacturing and commercial classes at the expense of the landed interest with whom power had traditionally resided.

The full force of this policy fell on those who had been major beneficiaries during the period of so-called 'High Farming' in the middle of the century, namely, the grain growers. In 1870-4 wheat had been selling at an average of £2.75 per quarter but by 1895-9 the price had fallen to £1.40 and it was simply quite impossible for the arable farmers of Eastern and South Eastern England to cultivate profitably at those prices. During the period from 1871 to 1901 the amount of land used for grain cultivation fell by 3 million acres.

The grain producers were facing competition on an intimidating scale. They had to endure a poorer climate, higher rents and soil which needed constant manuring in order to maintain its fertility whereas American farmers had a warm, sunny climate and vast tracts of virgin land which for many years would require little or no fertiliser. It is hardly surprising that British farmers were feeling the chill wind of this new competition and were demanding protection.

The difficult times being experienced by domestic grain producers benefited the country's livestock farmers. This was because the fall in grain prices in turn forced down the price of animal feed and thereby reduced the cost of producing meat. Additionally, this occurred at a time when improved methods of breeding livestock were producing animals that were characterised by higher yields of meat and proportionately less bone and which were ready for slaughter at a younger age than previous breeds. It was a useful bonus for livestock farmers that these events occurred when a general rise in living standards meant that consumers were able to afford to buy more meat.

Certainly the hard times being experienced by the grain farmers need to be put into context. They were only a fraction of all farmers and even before the Depression cereals had only constituted a small part of the total value of British agricultural output. In 1870-6 wheat, barley and oats had made up only 18.1% of the total value of Britain's agricultural output, and of this, wheat provided 8.4%. By 1911-13 the share of all cereals had declined to 10.3% and wheat to just 4.1%. By contrast meat, milk, wool, eggs, poultry and horses had made up 66% of the total value of output in 1870-6 and this rose to 75% by 1911-13.

Earlier historical accounts of the 'Great Depression' painted a picture of almost unrelieved gloom, this view being put over by Lord Ernle in his magisterial work, *English Farming Past and Present*, first published in 1912. He claims that the period 1879 to 1897 was uniformly disastrous for agriculture. However, much of the evidence he used was from the two Royal Commissions on Agricultural Depression of 1879-82 and 1894-7, the first of which had many members who were very influential large landowners with estates in the grain-growing districts of Southern and Eastern England. Ernle may have permitted the point of view of this vociferous but not necessarily typical lobby to make him draw a somewhat distorted picture of the state of the industry in this period.

While it was quite impossible for livestock farmers totally to disassociate themselves from the major changes that were taking place in world food supplies from 1870, they did have more options open to them than the grain-growing farmers. As Britain came to rely increasingly on cheap imports of factory-made cheese and butter after 1870, British dairy farmers frequently moved into the fresh milk trade where there was no likelihood of foreign competition. The virtual completion of Britain's extremely comprehensive railway system by the late 1880s and the development of a network of long-distance milk trains travelling at express passenger speeds greatly expanded access to fresh milk supplies and improved the health of Britain's now predominantly urban population.

Another sector of the industry which was to experience growth after 1870 was horticulture, taking advantage of rises in real wages and improved expectations which led to increased demand for fruit and vegetables. Near Peterborough, the Biggleswade and Sandy areas of Bedfordshire developed market-gardening on a substantial scale, Spalding made a name for itself with flower and bulb growing and the Fenland district around Wisbech, previously almost exclusively devoted to wheat, began to become a major centre for soft fruit growing.

CHAPTER FIVE

The Victorian and Edwardian Shows Held at Millfield 1887-1910

1887 was a year of great significance in British history because it witnessed a variety of ceremonies staged throughout the country to celebrate the Golden Jubilee of Queen Victoria's accession to the throne. For the citizens of Peterborough there was further cause for celebration because in that year the Agricultural Society moved to its new and larger Showground at Millfield near the well-loved windmill which stood at that time in Lincoln Road.

The local newspaper of 30 April 1887, reported this development under the headline, 'The Site of the Agricultural Show':

> *The Development of Peterborough goes on apace, and the old showground on the 'Rink Site', having been cut up for building purposes, the Peterborough Agricultural Society have this year had to find another site for the show. The site selected is the field known as Sexton Barns, between Westwood and Millfield.*

The use of the name 'Sexton Barns' for this new Showground seems at first glance to be very curious because historically and traditionally the 'Sexton Barns' of Peterborough were on a site which had been built over by the Great Northern Railway close to where the present railway station stands. Mr Taverner, who previously farmed the land now occupied by the Great Northern Railway, moved to Millfield and built new barns there which he called 'Sexton Barns', thereby unwittingly causing problems for future researchers. These buildings then quickly became referred to by all and sundry in Peterborough as the 'Sexton Barns'. They were part of the site which was purchased and developed to form the 'Millfield Showground'.

The Millfield site was described in the local paper, enthusiastically but perhaps with a little exaggeration, as 'picturesque', It consisted of two fields, the first of which contained a grandstand, a jumping course and a number of horse boxes. The second field was occupied by the judging rings and implement stands. There were two entrances to the Showground. One, for the visiting public, was opposite the Eight Bells Inn in Lincoln Road and the other, used by exhibitors and officials, faced the Millfield Post Office, which at that time stood just on the town side of the old windmill which had long been such a conspicuous landmark in that area. This mill, known locally as 'Adam's Mill', had six splendid sails and was, regrettably, demolished in 1937.

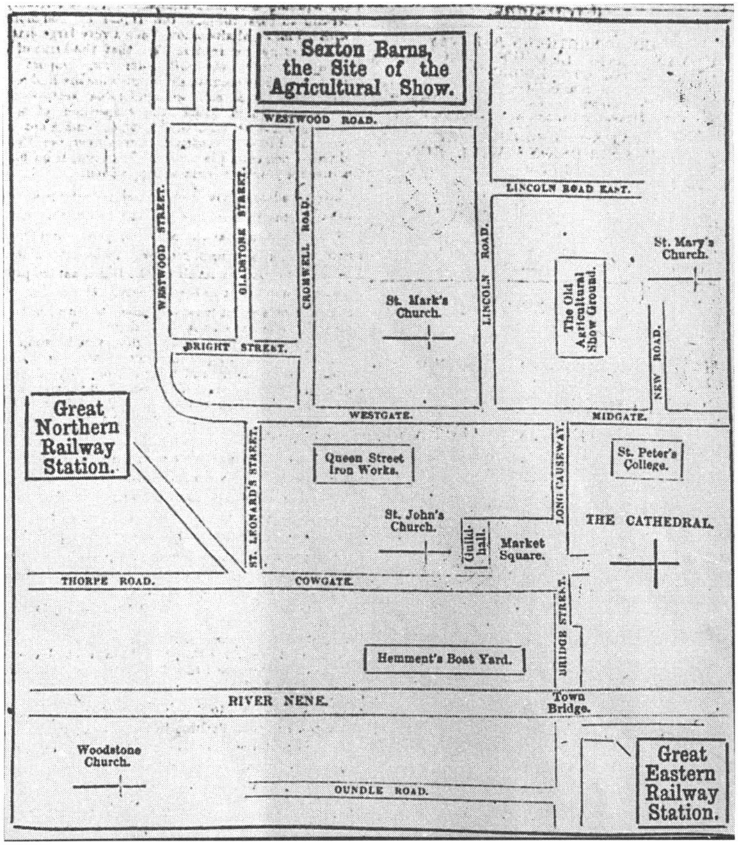

15. Location of Millfield Showground.

Published in the *Peterborough and Huntingdonshire Standard* of 30 April, 1887, this plan shows the position of the then new Millfield Showground in relation to other landmarks.

The *Peterborough Advertiser* waxed enthusiastically on its new Showground theme as it explained how the city was going en fete for the first Show at the new and improved site:

> *The City had more gay bunting than usual, and was in a state of animation throughout the day. The flys, waggonettes, horse buses and other vehicles plying to and from the Great Northern and Great Eastern railway stations went as fast as the horses could travel. The route from both stations, was through Park Road and Lincoln Road East to the show ground and via Lincoln Road, Wood Street, and St Leonards Street and Westgate on the return journey.*

Once more, visitors to the Show were treated to a fine display of fireworks. The main tableaux consisted of features on the subject of the Queen's Jubilee. In the agricultural machinery section, one item that attracted a great deal of interest was a vertical-boilered agricultural steam engine, then the largest ever constructed by the local firm of Barford and Perkins at their Queens Street Works and which was duly awarded a silver medal.

In 1888, the total number of entries ensured that in this respect it was the largest Show ever staged by the Society who must have been very glad that they had made the decision to move to the Millfield site. This provided nearly twice as much space as any of the Showgrounds previously owned by the Society. Particularly excellent this year were the equine exhibits and indeed experienced judges considered them the best ever. No fewer than 4,000 visitors paid at the gate on the second day with a further 1,440 paying in the evening just to see the firework show which was becoming a regular annual feature.

One especially interesting feature among the non-agricultural exhibits was a selection of the very latest bicycles provided by the local retailer, Rowe Bros. Healthy he-men and daring young damsels had taken to this radical new item of technology. In 1885 the Rover safety bicycle had been launched in Coventry by John Kemp Starley and had brought together the main elements of the modern machine. In 1888 a Belfast vet, John Dunlop, invented the pneumatic cycle tyre which proved instantly popular and thereby guaranteed success for the new cult of 'bicyclism'. Before this time, bicycles had often been made by village blacksmiths but they were heavy, cumbersome and difficult to manage. Now this lighter bicycle allowed young women of the middle classes some escape from chaperonage, despite the establishment a few years later of a Chaperon Cyclists' Association, and provided a very real stimulus for the equality of the sexes. This new craze went on to have its effects on women's fashions because long skirts were both inconvenient and dangerous when cycling and consequently a movement developed for 'rational' female clothing in the form of knickerbockers or divided skirts. For women, the bicycle was to set to become a powerful symbol of liberation, freedom and access to healthy recreation, particularly when the price of both new and second-hand machines came down in the 1890s. Evidence of local enthusiasm for the new form of transportation had been shown by the establishment of the Peterborough Amateur Bicycling Club in 1874 and its absorption of the earlier Tricycling Club in 1881.

For the Society, success followed on success. The Show was becoming so popular that it seemed a daring yet logical step to move to the staging of a three-day show and this was done for the first time in 1889 when, among other features, there was a revival of the poultry show which had no fewer than 500 exhibits listed in the programme, sufficient to ensure that it became a regular feature at Shows in the future. One innovation of interest was a show of harriers and beagles on the final day. No similar event occurred at any other Agricultural Show in the area and as many as fourteen packs of harriers and five packs of beagles were entered in this pioneering event.

Records both for the number of show entries and attendance were broken in 1890, 1891 and 1893 and the Society, ever mindful of the need to keep its facilities as up-to-date as possible, added some more space to the Showground by means of demolishing a number of small cottages adjacent to the windmill on

Lincoln Road. At the same time a new and more convenient entrance for exhibitors was established in Taverners Lane (now Taverners Road).

In 1894 all previous records for attendance were broken with a total of 10,600 visitors over the three days plus several thousand others who paid just to attend the firework display and dance on the final day. By now, the Peterborough Show had developed an enviable reputation on any number of counts but perhaps most of all for the quality of its equine displays, exhibits and competitions. These were generally reckoned to be of a standard equal to or even higher than those of the Royal Society's own shows.

During the 1894 Show, a presentation was made by the President, The Duke of Bedford, to Alderman William Barford, JP, of the eminent Peterborough agricultural machinery and implement manufacturers, Barford and Perkins. He was presented with a handsome silver Queen Anne tea and coffee service and additionally a gold chain and a silver bowl in recognition of his services to the

16. Fowler steam traction engine in Westgate, Peterborough.

This ponderous machine, usually hired in conjunction with ploughing apparatus, is seen making its way purposefully as part of a small convoy through the City Centre en route to the Millfield Showground. Many people thought these vehicles made one of the most wonderful smells known to mankind.

Peterborough Agricultural Society and the Foxhound Show Society. The presentation emphasised the close and beneficial working relationship that had developed between the Society and the elected members of the local authority. It also paid tribute to the long-term involvement of Barford and Perkins with the Society and its Shows as an exhibitor and competitor, its success in carrying off so many awards over the years and its long-term support for the Society, its activities and ideals.

In 1895 the Prince of Wales announced that he would be attending the second day of the Show, on Wednesday 10 July, and was particularly keen to support the Foxhound Society's display. It was decided to make this an occasion the City would remember.

17. Procession of steam traction engines in Lincoln Road.

The convoy, which was seen in illustration 16, is now pictured turning out of Westgate into the bottom end of Lincoln Road and pointing towards Millfield. Note the Masonic Hall between the two vehicles. This is still to be found at the junction of North Street and Lincoln Road. Note also the curious structure on the building in the top right hand corner. It is in fact a ventilator on a maltings which at one time came under the ownership of Allsopp's, one of the largest brewers in Burton-on-Trent. The exact date of illustrations 16 and 17 is unknown, but is probably in the early 1890's before the trams traversed Westgate and Lincoln Road.

The Prince and the Royal entourage arrived from London at the North Station aboard a train for Scotland which was driven as far as Peterborough by Sam Barley, a senior engine-driver who lived in the New England part of the City as did so many other railwaymen who worked for the Great Northern Railway Company. The Prince was met by Lord Burghley and they then transferred to an open landau, hauled by four magnificent greys and complete with postilions in fawn and gold livery. They made their way through streets lined with well-wishers to the Cathedral because the Prince had specifically requested the opportunity to view the famous west front. Having done so, he and his party moved on, again through large crowds, to the Showground at Millfield where they were greeted by the 6th Earl Fitzwilliam and escorted to the Royal marquee close to the main ring, where luncheon was served.

The Prince was driven in leisurely fashion around the various areas of the Showground in a horse-drawn carriage and of all the attractions on view, he showed particular interest in the large exhibition of farming implements. After his tour of the various activities, he and his party took their seats in the specially-built Royal Box in front of the main show ring where they watched the parade of prize-winning animals followed by an exhibition of high quality horse-jumping which was clearly of great interest to the Prince. The Royal Party left Millfield at 4pm, once again passing through streets crowded with onlookers but this time to the East station where a special train awaited them, this time driven by Driver Sheppard with Fireman Lambert. They were both from Peterborough and were based at the old Great Eastern Railway engine sheds which still stand in East Station Road and are now used as industrial premises. This train proceeded to Wolferton Station on the line between Kings Lynn and Hunstanton. This station, being situated close to Sandringham, was frequently used by the Royal Family. Indeed it had been extensively rebuilt with a range of luxurious amenities at the expense of the Prince of Wales himself.

The citizens of Peterborough had certainly entered into the spirit both of the Society's Show of which they were undoubtedly already very proud, and the visit of the heir-apparent. The City's streets were festooned with streamers and bunting, with many shops having special decorated displays, while across Narrow Bridge Street between the Angel Hotel and the Monarch Clothier shop was suspended an arch containing the emblem of the Prince of Wales Feathers and the words, 'A Hearty Greeting'. Perhaps most eye-catching of all, however, was an imposing triumphal arch, in a roughly Norman style, erected in Boroughbury near the Freemason's Hall at the corner of North Street.

After such momentous events as these, it was perhaps inevitable that the following year's show, although nearly equal in terms of visitors and the number of entries, was something of an anticlimax. In 1897, the year of Queen Victoria's Diamond Jubilee, the Show was most notable for an equestrian exhibition including water and fence jumping, a grand parade of riding and driving horses plus stirring martial music presented by the Robin Hood Rifle Band.

At its Annual Dinner held after the Show, The Marquis of Huntly gave a speech in which he made clear his view that the prosperity of the Peterborough area was

18. Decorative Arch at Boroughbury.

This mock fortified gateway was erected as part of the decorations put up to celebrate the visit of the Prince of Wales to the Peterborough Show in 1895. It spanned the bottom end of Lincoln Road near the Freemasons Hall at the junction with North Street. A more permanent masonry gateway stood for many years at the entrance to the Eastfield Showground.

very dependent on the success of agriculture and he considered that the endeavours of the Society were making a considerable contribution to ensuring a good future both for agriculture and for the City itself. The President for the year, P.A. Muntz, MP, alluded to similar sentiments in his speech when he stated that he had always looked with gratification upon the Peterborough Agricultural Society as one of the most successful and efficient in the Kingdom. Agricultural Societies, he told his audience, played a very important role in educating the agricultural mind.

In 1899, the *Peterborough Advertiser's* headline was: 'A Brilliant Success and a Record Attendance'. It then enlarged on this theme:

> *In its long and honourable career the Peterborough Agricultural Society has scored many distinguished successes at its Annual Shows–successes which have first built then solidified for it a reputation co-equal with that of the Royal Society. That reputation has been further strengthened by the 1899 Exhibition, the schedule of which comprised 94 classes of horses, cattle, sheep and pigs, compared with 88 in 1898 and in prize money the handsome sum of £1400*

Peterboro' Agricultural Society, 1903.

SHIRE STALLION

'Buscot Harold'

(16576)

Bay. Foaled 1896.

Sire—MARKEATON ROYAL HAROLD 15225.

Dam—13951 AUREA by THORNTON PREMIER 12551.

The Property of Sir A, HENDERSON, Bart., M.P.,

BUSCOT PARK, FARINGDON, BERKS.

HIRED BY

The Peterborough Agricultural Society

TO SERVE SHIRE MARES THE PROPERTY OF MEMBERS

Subscribing £1, at a Fee of £6 10s. each Mare.

All Fees must be paid at the Time of Nomination.

"BUSCOT HAROLD,"

Champion at the London Shire Horse Show 1898, 1899, 1900.

Aurea has won numerous Prizes, notably—1st Royal Counties' Show, 1893; 1st and Gold Medal, Royal Lancashire Show, 1896; 1st and Championship, Warwickshire Champion Shire Horse Show, 1898.

J. E. LITTLE,
Secretary.

Offices—COWGATE,
PETERBOROUGH.

19. 'Buscot Harold', 1903.

Stud card for the services of the Shire Stallion 'Buscot Harold'.

> *was offered. The entries were more plentiful than ever before in the Society's history with the exception of 1894, when a memorable record was established. This year they numbered 813 or 33 more than in 1898, and 127 more than in 1897.*

In the Shire Horse section, The Childwick Challenge Cup worth 20 guineas, donated by Sir J. Blundell Maple, MP, for the best shire mare or filly being the property of a tenant farmer, was won by Mr F.W. Griffin, of Boro Fen, with 'Grand Duchess'. This extremely fine animal was later sold for a new Peterborough record price at the Show to none other than Sir Blundell Maple himself. It is to be hoped that his munificence in providing the prize in this competition later proved to be a worthwhile business investment.

In 1900 the *Peterborough Advertiser* continued to print paeans of praise for the Society's efforts as exemplified by the following statement:

> *The Peterborough Agricultural Society breaks records with as much persistency as a modern cyclist. Every year the statistics of one department or another make a higher total than the previous 'best', and a new standard of comparison is hereby created. Twelve months ago it was the attendance which eclipsed the return of any previous Show; this year the entries followed suit, the aggregate of 1899 being topped by 104!*

The paper recorded that this was the 27th Show organised by Mr E.J. Little in his role as Secretary of the Society. Among interesting innovations at the Show was the 'Working Dairy' in which milking competitions took place and which was thrown open to dairy farmers from the Soke of Peterborough and Huntingdonshire. The implement display was once again impressive enough to attract a great deal of attention and potato growers were particularly interested in a potato harvester exhibited by Mr C.J. Moulton of March and Chatteris. This had been patented some years earlier but had been modified and was now found to be considerably lighter and therefore more practical.

In 1902 the Royal Agricultural Society, possibly due to the Coronation of King Edward VII in the June of that year, broke with precedent and decided to hold its Annual Show on the same days as those on which the Peterborough Society traditionally held its Show. Although the Royal's Show was staged that year as far away as Carlisle, it seems more than likely that both Shows were harmfully affected in terms of entries and attendances by this development. In the case of the Peterborough Show, livestock entries, for example, only numbered 715 and this was the lowest number since 1893. One interesting innovation at the 1902 Show was the inaugural appearance of a fully equipped ambulance tent.

In 1903 the Society approached King Edward VII to be its Patron. He accepted with alacrity and even went so far as to submit eight entries in a variety of classes at that year's Show.

Two events in 1903 reinforced the Peterborough Society's reputation for being involved with or reflecting up-to-date technological change. One of the exhibits at the Show was a novel electric motor van which could be used for all sorts of local light haulage work. The newspaper report stated that this vehicle could carry a load of 500 cwts, i.e., 25 tons! The authors believe this to be a mistake and that the van in fact could carry a payload of 50 cwts. It boasted a range of 40 miles before its batteries required recharging, possessed five different forward speeds and two reverse speeds and was fitted with electric lamps. It was displayed by the well-known local agricultural implement manufacturer and supplier, Hayes and Sons of Elton. Although little more seems to have been heard of this particular machine, a variety of battery-powered commercial vehicles made their appearance on the streets of Britain's towns and cities in the years to come and their successors can still be seen today, in the form of such things as milk floats. In retrospect, it can be seen that its inclusion among the exhibits at the Show was an indication that the days of the faithful horse as a prime mover were numbered, although this would certainly not have been obvious at the time.

The other notable event was the discussion that took place at the Annual Show Dinner about proposed legislation aimed at restricting the use of motor cars. The

20. The Angel Hotel.

This view is looking north along Narrow Bridge Street, towards the Market Place, showing the Angel Hotel on the left. The decorations are for the celebration of the Coronation of King Edward VII in 1902. The Angel was the scene of the Society's inaugural meeting and much of its subsequent business and social activities. It was a real landmark in Peterborough's City Centre and served both the business community and those wanting to socialise over a drink. From an historical point of view, it can be said that more than a building was lost when it was demolished around 1970.

21. Entrance to the Millfield Showground. Early 1900s.

This is a view in Lincoln Road, looking north and showing an entrance to the Millfield Showground and the 6-sail windmill. This area is about one mile from the City Centre and, when it was built, the mill's surroundings would have been almost entirely rural. The tram's destination was the Market Place and it was probably on a special service for visitors to the Show. The restricted width of Narrow Bridge Street meant that electric trams were unable to cross the City Centre and to head out across the Town Bridge to serve the southern suburbs of Peterborough such as Fletton, Stanground and Woodston. This put a serious limit on the usefulness of Peterborough's tram system.

Peterborough Advertiser reported that the Society's members found themselves on the horns of a dilemma. They could see the advantages that the motor car seemed to be able to offer but their loyalties were still very much with the horse and they had no wish to see the noisy, fast-moving motor car displace that noble animal from the urban streets and rural roads, lanes and fields. The high esteem and affection in which agriculturalists held the horse can clearly be understood in view of the increasing numbers and quality of equine entries at the Show. The shire horse stallion, 'Buscot Harold', chosen by the Society to serve the shire mares of its members is indicative of the quality of stock-breeding which the Society was in a position to encourage at that time.

At the 1904 Show, Barford and Perkins exhibited, for the first time in Peterborough, a hay press driven by a petrol engine recently introduced into this country. Another entry was an oil engine manufactured by Blackstones of

22. Queuing at Millfield.

Crowds waiting to enter the Millfield Showground around 1900. Note the thatched barn near the entrance which only serves to emphasise the rural nature of the area before Peterborough's outward expansion overtook it.

Stamford which had previously won the only medal awarded for implements at that year's Royal Show. Hayes and Sons were once again well to the fore.

In 1906 among the miscellaneous exhibits was yet another example of the wonders of new technology, an American 'Talking Machine'. Events had moved on from the Edison phonograph of 1878 which allowed its user simply to record his or her voice, to the invention by Emile Berliner of the flat disc which was designed specifically to allow its user to hear words or music that had been pre-recorded. With continuous improvements, these were already selling extensively in the USA but were still very much a rarity in this country. Just imagine how the crowds must have listened spellbound with amazement as the resonant tones of Caruso emerged from what appeared to be a small piece of wooden and metal machinery!

The report of the 1909 Show had the first definitive statement that the Peterborough Society was planning yet another change of location for their Showground. The *Peterborough Advertiser* declared that :

> *The increasing success which is attending the Peterborough Show is largely responsible for the acquisition of a larger site on the Eastfield Road and although next year's venue remains unaltered, in all probability in 1911 the trams loaded with visitors will be wending their way in the direction of Newark.*

23. A Family Occasion.

This picture dating from 1907 emphasises how the Show, while attracting the farming community, has also always attracted large numbers of people of all ages looking for a memorable and entertaining day out.

It was accepted that the Show of 1910 would be the last to be held at the Millfield Showground. This excellent site had provided a home for the Show for over 20 years during which there had been an almost continuous expansion of the livestock and other agriculture-related exhibits. The Society's success in attracting exhibitors and competitors and the public meant that the Millfield site was now becoming hopelessly congested. As with previous sites, it was being affected by housing developments nearby which were restricting the potential for continued expansion at this location. To ameliorate the situation, 24 acres of land at Eastfield were purchased by the Society, adjacent to the site of what is now Peterborough Regional College.

This last Millfield Show was of superlative quality. The entries for shirehorses, hunters and shorthorn cattle were especially outstanding and clear evidence that the Peterborough Show was building up an enviable record for its entries under these categories. In 1910 practically all the Royal and the Olympic champions appeared at the Show. In the Silver Challenge Cup for mares and Shire fillies, competition had never been more intense. If this cup was won twice in succession

24. Millfield Showground 1910.

This view looking across the Millfield Showground reveals the main Arena and also in the background the long-since demolished and very distinctive 6-sail windmill which was such a landmark on the northern fringe of the City at that time. This picture is also of particular interest because it depicts the last show held at Millfield.

or otherwise on three occasions by a member of the Society, the trophy, a magnificent piece of silver plate weighing 84 ozs and valued at £39, became his exclusive property. Mr F.W. Griffin had won the cup in 1907 and 1909 and he must therefore have fancied his chances and been severely mortified when the award went instead to Mr J. Bradley's yearling filly, 'Halstead Royal Duchess'.

At the Annual Show Dinner, the Mayor of Peterborough, Councillor W. Cliffe, J.P., in his response gave some interesting reminiscences of Peterborough and the Society and he expressed the following sentiments:

> *My first connection with the Agricultural Show was in 1848 when I was in the employ of Messrs Southam and Griffin, Ironmongers, whose shop in Narrow Bridge Street is now occupied by Mr Haydon. Their warehouse was in Albert Place beyond which there were no houses, only fields. Since then there has been a remarkable growth both of the City and of the Agricultural Society. The Show was then held in a small paddock up the yard of the 'Waggon and Horses'. They have since moved the Show to several localities and eventually to Millfield. It testifies to the growth and prosperity of the Society that even this commodious ground is too small for them and next year they will be moving to an even larger ground. Agriculture was of course the first pursuit of man. Adam having the Garden of Eden to till and being turned out because he sold the fruit. From that day to the present, agriculture has continued to flourish...*

So the valedictions were expressed for Millfield and the Showground there closed its doors for the last time. The Peterborough Agricultural Society however, continued its onward march and a new chapter in its history was about to begin.

British Agriculture and 'The Great Depression'

The successes of the Peterborough Agricultural Society which had brought it to the stage of outgrowing yet another Showground are all the more remarkable given the general situation facing agriculture from the mid-1880s to the outbreak of World War One. While today's historians tend to call into question whether the situation in agriculture really warrants the description 'The Great Depression', that there were deep-rooted problems can hardly be questioned and this was recognised by the fact that agriculture was the first industry to be given, in 1889, a government department of its own. Agriculture is an industry which encompasses a wide range of products and methods of production and the various components of the industry were by no means equally or simultaneously affected by the growth of competition from abroad. Indeed many parts of the industry benefited from the increasing population of the period and general rise in real wages which meant that there was buoyant demand for many meat and dairy products and for fruit and vegetables grown in the UK.

For a time, some farmers enjoyed reductions in the cost of fodder for their livestock which more than compensated for any reduction in domestic demand for their meat as a result of imports. Later, however they rapidly became disillusioned in the 1890s when major developments in cold-storage methods in international transport turned a flow of imported meat into a veritable flood which started to damage the pastoral farmers of the west and north just as the cereal or grain farmers of eastern England had been hurt some years earlier.

Reading through the debates in *Hansard*, the published verbatim report of Parliament's deliberations, would lead one to suppose that the big landowners of eastern England had entered a state of terminal decline as masses of cheap grain from overseas poured into this country. This importation of overseas produce had, not surprisingly, started with those products that were easiest to transport, namely wool and wheat. Wool prices for domestic producers were falling up to 1884 because of almost continuous annual rises in imported wool. This was no mere bagatelle for farmers in the eastern counties for sheep at that time had an important subsidiary role to play in the pattern of arable farming in the region.

The situation as far as wheat imports were concerned was very serious. By 1914 Britain was importing more than 5 million tons which was about three times the total amount of domestic output. The British farmer was hanging on with desperation to around 25% of the home market but in so doing was accepting prices that much of the time were below 30s a quarter, as compared with 45s per quarter around 1880. A quarter was the equivalent of 4.5 cwt. What compounded the difficulties was that other cereals such as barley and oats had already flooded Britain's markets ahead of wheat and there were therefore no alternatives for the wheat grower to turn to when foreign wheat started being imported in such huge quantities. Through the unfavourable weather between 1883 and 1890, grain farmers clung desperately to the hope that things would turn round if only they had a run of average seasons. On occasion radical improvement was brought about by a change of tenant, particularly to Scots farmers who brought their economical dairy farming practices and applied them with some success to the clay soils of East Anglia. It is hardly surprising that land rents fell in this period as landlords were anxious to retain their tenants. The situation then, was a complex one but certainly as far as the grain producers of the eastern and southern counties were concerned, the general trend was one of falling prices and profitability.

For all that, there were some increases in productivity at this time. Pirie's plough had first appeared in 1863 but was only now coming into general use and it allowed one man with two or three horses to replace two men with four horses. The self-binding reaper had also come into common use by the mid-1880s while minor improvements in threshing tackle allowed work to be done more quickly and left the grain cleaner. As far as grassland husbandry was concerned, there were new mowers, horse-rakes, swathe-turners and elevators. At the same time, advances in metallurgy were assisting farmers: better ironwork was making a wide range of implements far more effective, wire was being increasingly widely used in fencing and the Gilchrist-Thomas process in steelmaking was responsible for a by-product, basic slag, which after 1896 came into general use as a fertilizer for grassland.

The move towards specialised farming, especially of vegetables, fruit, potatoes and most of all, milk, continued after 1900. Though the number of sheep went down, cattle numbers increased slightly while the quality of the animals rose as a result of the ongoing work of the Breed Societies. More meat was being consumed and home producers made profits from selling meat which was of higher quality than any being imported. At the same time, increasing amounts of milk were being consumed in the towns, the railways carefully fostering what for them was a specialised and highly lucrative traffic. By 1908, no less than 20% of the value of UK farming output was dairy produce. While the labour force of agriculture as a whole was continually falling, that proportion involved in the market gardening sector actually increased. As compared with 1881, the census of 1911 showed a 25% fall of employment in agriculture, but market gardening employment had increased by no less than 77%.

The years leading up to the First World War were characterised by the return of unhappy labour relations within agriculture. One contributory factor was the absence of any effective form of trade union for farm workers after the largely unsuccessful efforts of Joseph Arch in the 1870s, and clear evidence of this came with the widespread evictions of labourers from tied cottages after the Liberal Party gained its landslide victory in the general election of 1906. Most of those workers sacked were dismissed because it was felt that they were radicals or simply because it seemed that they had defied their employers' wishes and had rather indiscreetly boasted of voting Liberal.

Once again East Anglia was at the centre of militant activity as it had been back

25. Barford and Perkins hay binding machine.

This picture shows a working example of the products of Messrs Barford and Perkins, the well-known Peterborough manufacturers of agricultural machinery, who helped establish the City's reputation as an engineering centre. c.1910.

in the 1870s. Closely in alliance with the Liberal Party, the Eastern Counties Agricultural Labourers' and Small Holders' Union was established in 1906. 1910 was a year of enormous political and labour unrest and there were some particularly acrimonious strikes in Norfolk in which the workers were defeated and victimisation followed. For their part many of the strikers were convinced that their supposed allies in the Liberal Party had let them down and, significantly, at its 1911 conference, the union voted for affiliation to the Labour Party and the TUC. In 1912, the Union, whose membership was still very much localised, decided to change its name to the National Agricultural Labourers and Rural Workers Union. In 1920 it was renamed again, this time as the more familiar National Union of Agricultural Workers. Labour relations continued to deteriorate through the years 1911 to 1913 and strikes were frequent, not only in East Anglia but in Bedfordshire, Northamptonshire and elsewhere. Many of these were successful from the workers' point of view because their bargaining power was enhanced by serious labour shortages in many parts of the country and also because there was a slow return to prosperity.

It is perhaps unsurprising that the development in collective organisation among the workers' side of the industry was paralleled among the employers. In 1908 the Lincolnshire Farmers' Union was a response to the growing tide of labour unrest in that county. Its growth proved to be meteoric and it quickly spread across the country, changing its name to the National Farmers' Union (NFU) and soon becoming recognised and valued as the leading mouthpiece representing tenant farmers' interests as opposed to those of labour on the one hand and the landowners on the other. It is worth noting that Lincolnshire was also the county where the Central Land Association, later the Country Landowners' Association had been founded in 1908. This became the acknowledged body lobbying on behalf of the big landowners.

Between 1900 and 1914 there was little change in the acreage under grain, only barley declining. This was a reflection of the marked fall in the country's consumption of beer, a perhaps rather unexpected result of an overall if unevenly distributed rise in living standards. During this period agricultural engineers continued to exercise their ingenuity in finding ways of helping farmers to maintain their acreage with a reduced workforce. Concave mouldboards and more durable steels improved the effectiveness of ploughs and reduced the physical effort required by man and beast. Mechanical rick-lifters were improved as were corn-drills. Small stationary oil engines began to be deployed for a variety of jobs around the barns and in the farmyard.

In assisting these advances, there were two basic stages, research and development and the diffusion of new knowledge and practices. As regards the former, the industry had previously relied largely on the Rothamsted Research Station in Hertfordshire but in the 1900s, Cambridge University began to be involved with animal nutrition and plant breeding while Manchester and Birmingham researched agricultural zoology and the new University College at Reading began to specialise in dairying.

The new farming knowledge was diffused through a variety of channels. Government finance supported about a dozen agricultural colleges and several County Councils established advisory services, sometimes working with the agricultural colleges to arrange demonstrations to interested farmers. There were also several established and widely circulated farmers' periodicals. However probably the most important of all were the big Annual Shows of the Royal

Agricultural Society and the best of the county and local Agricultural Societies' Shows. These Shows effectively disseminated news of developments and innovations elsewhere and they were equally effective in advertising the multifarious products of the agricultural engineers who were so keen to use them for promotional purposes.

The last Show at Millfield in many ways epitomised the state of British agriculture on the eve of the First World War. Grain farming after its long period in the doldrums appeared to have achieved something of a comeback. Animal husbandry was doing well in the areas of meat and milk while the horses, cattle, pigs and sheep in the Main and various minor rings would be the focus of much attention. A wide variety of products would be found along the sidewalks of the Show. Small stationary engines would be merrily phut-phutting away in various parts of the ground. Elsewhere the scientist, albeit unobtrusively, was lurking in spirit if not in body behind the products on the stands displaying the wares of the seedsmen and the fertiliser and pesticide manufacturers. On the face of it, all appeared to be bustling. A somewhat discordant presence, however, was that created by the advertisements of the Canadian and Australian migration departments, using sweet words in an attempt to lure away those young ploughmen and cowherds who had not already drifted off the land in search of the jobs and bright lights of the towns. We know that some of the young men who did indeed migrate to the countries of the Empire actually returned to the mother country and joined the armed services alongside their school friends and their relations. Sadly, many of these young men did not survive to return to their adopted countries.

The period leading up to the First World War marked the closing phases of a rural society which, at least in its social structure, had shown a remarkable degree of apparent stability over a period of two hundred or more years. It was a society of tripartite form with the landed aristocracy and gentry who owned the greater part of the land, the tenantry who farmed it, and then the substantial, although declining wage-earning workforce who laboured on it. To these three basic groups could be added the clergy, a range of professional people such as land agents, craftsmen, tradesmen and those workers engaged in cottage industries and other activities closely allied to agriculture such as malting, brewing and basket-making.

The wider economic and social processes from the late eighteenth century onwards, such as the huge growth in population, the development of modern manufacturing industry and the appearance of large urban agglomerations of people, industry and commerce set in motion a number of major stresses and strains in rural society. The large landowners found their social and political domination challenged by an urban bourgeoisie with industrial and commercial interests. Over the nineteenth century as the status of land-ownership declined and rental income fell, many landowners sold their estates which were then broken up while the families themselves were dispersed. Hence the squire and his family whose ancestors had dominated a village for centuries might take themselves away, never to return. This inevitably left a vacuum.

The upshot of this was that farming became an ancillary activity in an economy dominated by manufacturing, by commerce, trade and finance which meant that those farmers who remained had to become more commercially-orientated, more open to the application of scientific and technological know-how and therefore more prepared to use their capital intensively. The old-style farmer who believed in and practised the concept of custodianship of the land with responsibility both to the countryside and to its workforce, largely gave way to the capitalist farmer with little time for sentiment.

Labourers and their families were inevitably affected by the growth of industry, the expansion of the towns and the intrusion of cheap food. By the early 1880s cheap imported food was helping them to eke out a living from their low wages but for many the future lay with the flight from the land as they listened to the honeyed coaxing tones of advertisers extolling the benefits of industrial employment or emigration. Those who left the countryside were often the youngest, most intelligent and vigorous. The pool of available agricultural labour went down and although the wages of those who remained often rose because of scarcity value, frequently they were the older or the less able workers. In practice it was often the womenfolk who were most attracted to the idea of a new life in the towns, coveting such luxuries as piped water and other services, the possibly better housing, the wider range of shops, the abundance of jobs in domestic service or in industry and what were perceived as better prospects for the children.

Education had become compulsory in the 1880s at about the same time that large numbers of rural male householders obtained the vote for the first time. This meant that the farmworker was now somebody to be wooed both by the trade unions and the political parties. While education was supposed to be the key to making democracy meaningful, an informed assessment of weighty political matters was something which cut little ice with many rural labourers. They remained complacently ignorant of the world outside their immediate sphere and thereby gained a reputation for an almost bovine stolidity and ignorance, the suggestion being that all those with anything about them had already got off the land. This was somewhat unfair because while many farmworkers were indeed only semi-literate, they were increasingly being called upon to master a wide range of skills for their everyday work and these were often much wider-ranging and demanding than those required by the largely unskilled but better paid factory operatives of the manufacturing towns.

It was inevitable that the nineteenth century saw a great many traditions, practices and superstitions disappearing, redolent of the pre-industrial countryside Local habits and amusements, local dishes and local dialects were all tending to die out and feasts, fairs, wakes and other long-standing customs were vanishing with great rapidity. Even as late as the end of the nineteenth century in south Lincolnshire, the men on Plough Monday, the traditional symbolic start of the new ploughing year, still carried a coulter, this being the iron cutter in front of a ploughshare, decorated with ribbons and then went round the village begging for a supper, as their forbears had done for centuries. The old style of village life was starting to decay and the village itself was changing its appearance.

Increasing numbers of abandoned cottages were to be found as were derelict windmills, symbols of the decline of rural industry. Silence reigned in the closed workshops which had once resounded to the variety of noises accompanying the daily labours of highly skilled village craftsmen such as blacksmiths, carpenters, masons, coachmakers and wheelwrights.

The character of the village and the tenor of rural life was changed permanently as the countryside became more and more permeated by urban and industrial influences. However the changes that occurred to the countryside and to the lives of those who lived in it before the First World War, were no more than a prelude to the far greater change that was to follow in the years after hostilities ceased.

CHAPTER SIX

The Shows at Eastfield 1911-1965

The Society's new Showground at Eastfield on the northern edge of the ever-expanding City was opened by the President of the time, Lord Rothschild. This land had been purchased outright by the Society from Mr S. Vergette whose house was on land close by and it consisted of just over 24 acres. It was situated close to the small eminence of Newark Hill and had the great advantage from the Society's point of view of being nearly ten acres larger than the old Millfield site. An eye-catching and substantial entrance archway was built, fronting onto Eastfield Road and the site enjoyed easy access to and from the City Centre using the trams of the Peterborough Electric Traction Company.

Among the notable entries at the 1911 Show were four from King George V. These were 'Golden Roch II', a shire stallion which had gained honours at an earlier Royal Agricultural Society Show, a shorthorn cow called 'Lancastrian Gayety' and two heifer calves, going by the curious names of 'Watercress' and 'Orange Robe'. These latter animals did the King proud and carried off the top prizes. The total number of livestock entries was 786, a favourable figure compared with the last Show held the previous year at Millfield and this gave the organisers reason to be optimistic that the transition from the old to the new Showground had been accomplished without any great loss of support from competitors and visiting public. One of the most prolific of the local competitors at the Show was a name which turned up with almost monotonous regularity year after year among the lists of participants. This was Mr F.W. Griffin, clearly a man who liked to be noticed for he excelled himself with no fewer than 33 entries at the 1911 show compared with a mere 24 the previous year at Millfield. He led the field of entrants by some distance from Messrs Henson and Son of Eastgate with 17 and Mr F.Gee of Thorney and Mr R. Nutt of Peterborough who tied with 11 entries each.

One of the largest non-agricultural displays at the Show was a stand advertising the splendid opportunities available for those who decided to emigrate to Rhodesia. This colony must have been in desperate need of immigrants from the mother-country because a through fare from England to any railway station in Rhodesia was available for just £11! It is possible that the journey by train and boat and taking a tedious 21 days may have put some people off. It would be intriguing to establish just how many people were tempted into seeking a golden future for themselves and their families as a result of viewing this display at the Show.

26. The dare-devil pioneer air-ace Frank Goodden.

Goodden can be seen flying his Moraine-Saulnier monoplane and thrilling the crowds with his low flying at the 1914 Peterborough Show. His speciality was looping the loop and he held the record for the greatest number of loops performed in one flight - 14. He performed this intrepid feat in a Caudron flying machine.

The 1913 Show saw the introduction of a competition involving old stage coaches and this proved very attractive to the paying public and was the first of several such events over the ensuing years. The competition had sponsorship from Mr Stanley Brotherhood of the well-known local engineering company, himself a great coaching enthusiast.

In 1914 the outstanding feature of the Show was a display of daredevil air stunts by the pioneer flyer, Frank Goodden. How the crowd must have stood engrossed with a heady mixture of terror, amazement and admiration as he looped the loop and performed various other acts of aerial derring-do in his Moraine-Saulnier monoplane. For some this would have been the first time that they had ever seen an aeroplane and the Show therefore maintained its reputation for being the place in Peterborough where the inventions that would impact on tomorrow's world, were first seen in the City. The newspaper report of his display records:

Soon after four o'clock the aeroplane left the hangar to the strains of Donald Fraser's bagpipes. The engine was started and the aircraft rose like a bird. Round and round it soared until an altitude of about a thousand feet was reached. Racing against the wind towards the village of Eye, the thousands of spectators gasped as they saw the aeroplane suddenly rise into a vertical position, then fall backwards, with its landing wheel above, and the airman hanging in his straps below. For a breathless moment the aeroplane thus glided on its back, the

> *engine silent. Then with a swoop, the circle was completed. Four times this was repeated. Some Peterborough people may have seen good flying before, but never such wonderful banking. Goodden simply waltzed around a pivot, as if there were no such peril as side slip. He would be racing against the wind one moment. The next, he had performed half a somersault, and was vol-planing down wind in a manner which would have been incredible a year ago.*

He repeated the display several times during the evening and took daring passengers up for flights. The show ticket holder who won the draw for a free flight was Mr A. Turner of Fletton near Peterborough and it would have been rewarding to record his responses not only when he heard that he had the winning ticket but also later, on having successfully concluded his flight. Incidentally the Scottish piper Donald Fraser, whose musical efforts had dramatically accompanied the emergence from the temporary hangar of Goodden's aircraft, was a celebrity himself, being a favourite of the late King Edward VII, for whom he had often performed with the bagpipes as dawn rose over the Royal retreat of Balmoral.

Just before the 1914 Show, Mr J.E. Little who was serving his fortieth continuous year as Secretary to the Society, told the local press:

> *The popular innovation of last year provided by the 'Coaching Competition', will again furnish a great attraction with ten coaches competing and we are assured of a very fine spectacle.*

Another innovation at the 1914 Show was 'Musical Chairs on Horseback', a spectacle which provided some excitement and much humour for the attending crowds then and which is still popular at such events as gymkhanas. It was based on the popular parlour game of the time. To start with the chairs were distributed in a fairly random way around the main ring. The band then struck up only to stop abruptly which was the cue for a frenzied rush of riders, jumping off their horses and dashing to the nearest seat. Each time the music stopped there was one less chair in the ring and eventually, of course, there was only one rider left and he or she was clearly the winner. This kind of spectacle may have provided some entertainment for the crowds and helped to swell the numbers who paid their entrance money at the gates but it was reported that there were some among the old stalwarts of the Society who disapproved of this kind of frivolity. They felt that the Show's prime purpose was the encouragement and dissemination of good practice in the farming community and that those features of the show which could loosely be described as entertainment were to some extent in conflict with this purpose. Such arguments were by no means restricted to the Peterborough Agricultural Society and indeed they continue to be heard to this day.

This last Show before the Great War was regarded as a considerable success. The total number of livestock entries was somewhat down on the previous year but was undoubtedly affected by the outbreak of foot-and-mouth disease which prevented any entries in the sheep classes. At this Show Mr Joseph Mackinder broke all previous records by officiating at both the judging and organising of the shire horse entries for a remarkable 44th successive year. A further notable achievement was that of Mr Charles Tuppin, who beat all comers in the coach horn blowing championship yet again. Considering that he had performed the feat 24 times before, this was a remarkable achievement but it was in danger of becoming distinctly monotonous. His virtuosity on the coach horn had already won him the accolade of World Champion so perhaps a trip to the Peterborough Show was small beer for a man of his calibre. The total attendance at the Show was 11,903 and on the strength of this the Society was encouraged to go on to plan an even larger programme of events for 1915. However their aspirations were jeopardised by the assassination of the Archduke Ferdinand at Sarajevo, the event which precipitated the barbaric carnage of the First World War.

In November 1914 the Society held a meeting to decide whether to abandon its plans for a Show in 1915. There was a widespread feeling that the war would soon be over and the November meeting ended with the Committee avoiding making a decision on the next year's Show, the feeling generally being one of 'wait and see'.

In April 1915 a special meeting took place in the Society's normal venue, the Angel Hotel, situated where the large shop occupied by W.H. Smith now stands in Bridge Street. The Angel was for so many years a distinctive and well-loved City-centre landmark and meeting place that its demolition was widely considered to be a particularly wanton piece of post-World War 2 vandalism masquerading as town-planning. After considerable discussion the members resolved that:

> *The holding of a three day's show on the usual lines was impractical and undesirable: impractical because, owing to the emergencies of a state of war, the railways might not be able to supply the usual facilities for the delivery and return of stock. The usual holiday and pleasure side of the show would also be badly affected.*

A motion to abandon the show for the year was duly proposed by the 7th Earl Fitzwilliam and was carried by 35 votes to 14. The war lasted until November 1918 and it was not until 1919 that the Society was able to stage its normal Show once more. Something of a compromise was reached with the decision at the April 1915 meeting that one-day shows for horses in March and mares and foals in September would continue for the duration of the war, however long that might prove to be.

British Agriculture and the Great War.

The coming of war saw the troubled relationship between employers and workers being set aside for its duration. The state intervened in all aspects of the economy, not least in agriculture and, perhaps ironically, this had the effect of assisting the growth and the influence of the National Farmers' Union, the Country Landowners' Association and the agricultural trade unions. This was because all three were required to be actively involved in official policy and decision making.

The main changes affecting British farmers and their incomes during hostilities were, of course, a substantial reduction in foreign imports and a large rise in domestic demand consequent on the temporary wartime phenomenon of full employment and well-filled wage packets for those whose employment serviced the requirements of the war. The Government lumbered along, making policy up as it went, but its actions ensured that by 1917, agriculture had become an industry acting under close and strict government control. The marketing of agricultural produce was effectively a government monopoly. Price controls assisted consumers while guaranteed prices gave succour to the farmers.

Few people had believed in 1914 that the war would last long and this dangerous misreading of the situation combined with its traditional reluctance to take pre-emptive or swiftly responsive action, meant that the Government took a long time placing agriculture on a war footing. Even in late 1914, large numbers of agricultural workers were still unwisely being allowed to enlist in the armed services thereby seriously depleting the available labour force on the land while the Army was requisitioning thousands of horses together with a large tonnage of fodder. Additionally, reduced imports of foreign foodstuffs and fertilisers were placing British farmers under a great deal of pressure. Consumers in turn were being pressurised as retail food prices rose 30% by the summer of 1915 and a further 30% by June 1916 over the levels pertaining at the beginning of the war.

The loss of around 250,000 members to the armed forces had hit the trade unions very hard but their position was definitely helped by the Corn Production Act which required them to be represented on the new Agricultural Wages Board and the local District Wages Committees. So it was 80 years after the courageous, pioneering efforts of the 'Tolpuddle Martyrs' that the collective bargaining role of agricultural trade unions was officially recognised by the State.

The return to normality

The ending of the war was met with extreme relief and the wave of euphoria was reflected in the headlines with which the *Peterborough Standard* greeted the 1919 Show, the first full show for five years. Headlines described it as a 'A Wonderful Show', proclaimed enthusiastically that 'The King was a prize-winner' and spoke exuberantly about the new records that were achieved when 20,000 visitors were recorded as passing through the turnstiles for the first time.

Among the more notable non-agricultural events on the third day of the 1919 Show, was the appearance above the Showground of a dirigible airship. The inhabitants of the Peterborough area had seen many of the dreaded Zeppelins in the war but fortunately this apparition had no hostile intent. It was none other than the R33, which first appeared from the Crowland direction and spent some time almost stationary above the Showground before moving slowly away into the cloud and mist. There was intense interest in aviation at that particular time which is not surprising because it was the year in which Alcock and Brown, flying a Vickers-Vimy Bomber, had made the first direct non-stop powered crossing by an aircraft of the Atlantic Ocean from Newfoundland to Ireland and around the same time, the R33's close sister, the R34, had performed its historic transatlantic return flight from Britain to the east coast of the USA. The R34 may well also have passed over Peterborough when it came back from Long Island because its

UK destination was Pulham in Norfolk. These airships were most certainly eye-catching, measuring around 670 feet in length and having a diameter of 80 feet. Their engines, five in number, generated a total of 1,250 horsepower. They could be used for either civil or military purposes, were designed to cross the Atlantic without stopping and had a maximum speed of 70 mph.

The cattle classes of the 1919 Show were considered to be of particularly high quality and Mr J.R. (Ray) Horrell of Longthorpe achieved a notable double winning the Champion Shorthorn Dairy Class with 'Westwood Dutch Girl', bought at the Peterborough cattle market and 'Westwood Dairy Maid' which carried off the trophy in the tenant Farmer Milking Class. King George V must have been gratified to have won the first prize in the Shire Yearling Filly Class.

The A.G.M. of the Society that year was held on the 3 December and was notable because the local newspapers reported that it was the scene of something of a palace coup d'etat. First the existing Chairman resigned, ostensibly on the grounds that he felt new blood should be brought in but probably because he was

27. Leading Lights in Peterborough's farming community.

In front of the NFU stand at the 1921 Show are some of the pillars of the area's agricultural community. In the back row, third from left is a youthful Robert Bibby, the last Secretary of the Peterborough Agricultural Society to combine the job with that of Secretary of the local branch of the National Farmers Union.

being pressured to do so. Then, in an extraordinary move, James Little, the Society's Secretary, was asked to leave the room while the members discussed his future. When he was called back in, it was to be told that the Society had decided to dispense with his services, also on the grounds that new blood was needed. It was recorded that Mr Little reluctantly agreed to retire but the whole occasion seems to be a very churlish and shabby way to treat a man who had given the Society 45 years of continuous service. It seems even more unreasonable because the minutes of the previous A.G.M. contained an item in which the Council expressed the hope that Mr Little would complete his 50 years in harness as Secretary. James Little was succeeded by Mr T.B. Turner whose stay was a short-lived one however, because in the teeth of much competition, he was successful in gaining the coveted position of Secretary to the Royal Agricultural Society of England.

In 1920 while the Society's annual Show was taking place, the Broadway Kinema showed a film of the Peterborough Show of 1911 which had, of course, been the first Show to be held at Eastfield. If this film had only survived it would have provided later generations with an absolutely invaluable visual record of the Shows of that period. Regrettably, as so often happens with valuable archive material, this film has simply disappeared as indeed has much other material that would shed more light on Peterborough's past in general and the Agricultural Society's Shows in particular.

The 1921 Show was perhaps most notable because it was the first under the stewardship of the Society's new Secretary, Mr Robert Bibby. He quickly demonstrated his organising abilities and was helped by the fact that the weather was excellent over all three days of the Show.

In 1922 the *Peterborough Standard* as usual reported the Show in great detail and was especially complimentary about the standard of the Shire Horse classes. It described the many outstanding entries at the Show and went on to comment:

> *It should be stated first of all in speaking of the Shire section that the exhibition as a whole was the best ever seen at Peterborough. This Show ranks in a place of its own among Shire Horse Breeders. Time and time again decisions made at the Royal Society Show have been reversed here and this Show was no exception to the rule.*

The Brood Mares provided the highlight among a very strong entry in the equestrian classes and the splendid 'Pendley Lady' was undisputed victor. She was owned by Col. V.D.S. Williams, father of Dorian Williams, later to gain fame as a show-jumper and media commentator. The well-known local competitor F.W. Griffin was among the winners yet again while in the Open Colt Foal Class, Messrs Bishop and Measures won with 'Champion Goalkeeper'. Mr W.R. Odam's 'Field Marshall' was a very close second. In the Implements Class, the prominent local firm of Peter Brotherhoods attracted much interest with their 'Peterbro' Agricultural Tractor.

Earlier in that year, the Government had repealed the wartime measures with which it had guaranteed prices and implemented wage controls. The result was an almost instant deterioration of conditions in the industry. Prices tumbled and

28. The 'Peterbro' Tractor.

Dated 1922 this is an advertisement for the 'Peterbro' tractor produced by Peter Brotherhood, a company still active in high quality engineering within the City of Peterborough although not currently involved in the manufacture of tractors.

farmers shed labour and cut wages, or both, the unions proving unable to stop or contain this process.

In the early spring of 1923, matters came to a head in parts of north and west

Norfolk with a localised but bitterly fought strike in which new tactics were employed. Flying pickets in the shape of striking labourers peddled furiously on their bicycles around the normally tranquil Norfolk countryside trying to prevent non-union labour from being used for seed-drilling. Large numbers of police had to be drafted into the area and at one point an exasperated farmer loosed off his crow gun at mocking pickets. The strikers replied with stirring refrains from the 'Red Flag' and 'Onward Christian Soldiers', a curious and paradoxical pairing redolent at the same time both of class struggle and evangelical Christian fervour. This confrontation continued to rumble on for some time and was only settled by the election into office of the first Labour Government in April which cobbled together a settlement which did something to save the Union's face. One of the men's most bitter grievances was the use of casual labour and this was now banned. However the strike had crippled the Agricultural Workers Union which was powerless to prevent widespread sackings and evictions from tied cottages. This proved to be the last strike of national significance in British agriculture and it was evident that the way forward for the union was full participation in statutory organisations like a wages board.

Without the protection of supported prices, it soon became very apparent that depressed conditions were returning for the industry. Once more it was the arable sector that was hardest hit. This had been encouraged to expand and intensify production during the War and the farmers of East Anglia were again among those who bore the brunt of the unfavourable conditions. Farmers looked for some kind of continuation of government direction and assistance but free trade and laissez-faire were still the watchwords. In 1926 a White Paper was published and this strongly urged farmers once again to switch from cereals to dairy and livestock production encouraged by the cheaper feedstuffs then available which in turn were created by falling cereal prices.

Various minor measures affecting agriculture were introduced by the governments of the 1920s, their main thrust being an attempt to limit unemployment in the rural areas rather than to make any full-blooded intervention in the industry itself. For example, the provision of credits was allowed for under the Agricultural Credits Act of 1923, money was made available for land drainage and, of particular interest to farmers in the fens and eastern counties, the British Sugar (Subsidy) Act of 1925 which gave a subsidy for sugar beet production for ten years. The result of this Act was that by 1929 eighteen factories were operating in the U.K. and 230,000 acres of farmland were under cultivation for the production of sugar beet. One of these factories was at Peterborough. It was opened in 1926 on land adjoining Oundle Road in Woodston. It has since been closed as part of the British Sugar plc's policy of concentrating production in a small number of very large plants.

On 5 July, 1923, Mr James Golby Barford died, the surviving son of Alderman William Barford, head of the firm of Barford and Perkins, the engineering company who for so many years had conducted their operations from the Queen Street Ironworks. The site has disappeared, being completely subsumed within the City's Queensgate Shopping Centre, but this company may well have been the

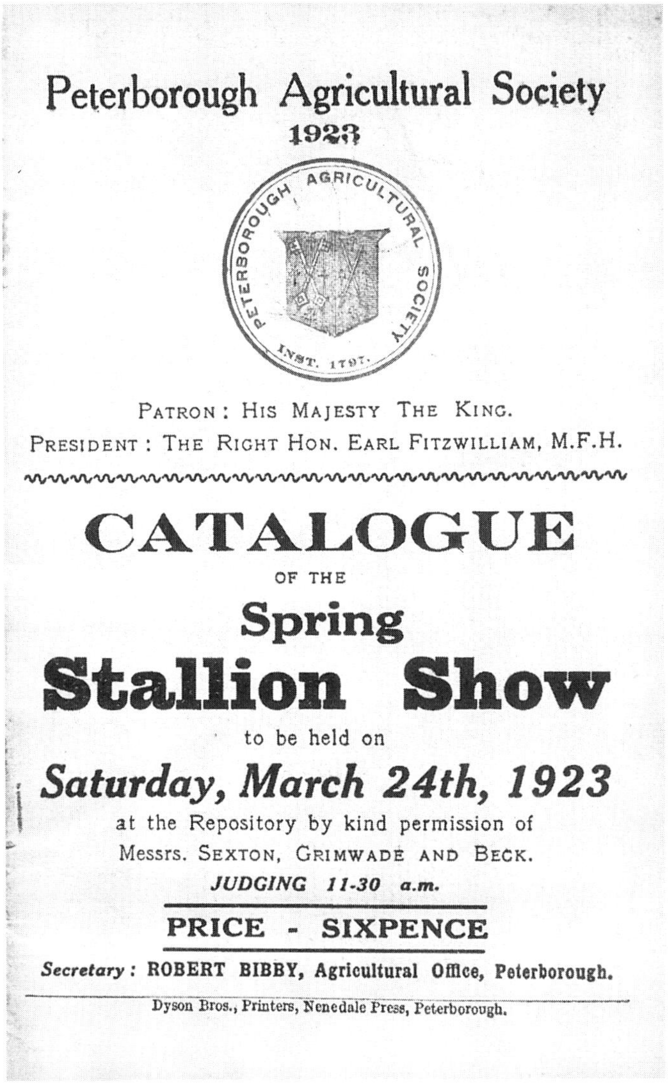
Peterborough Agricultural Society
1923

PETERBOROUGH AGRICULTURAL SOCIETY
INST. 1797.

PATRON: HIS MAJESTY THE KING.
PRESIDENT: THE RIGHT HON. EARL FITZWILLIAM, M.F.H.

CATALOGUE
OF THE
Spring
Stallion Show
to be held on
Saturday, March 24th, 1923
at the Repository by kind permission of
Messrs. SEXTON, GRIMWADE AND BECK.
JUDGING 11-30 a.m.

PRICE - SIXPENCE

Secretary: **ROBERT BIBBY, Agricultural Office, Peterborough.**

Dyson Bros., Printers, Nenedale Press, Peterborough.

29. Catalogue of the Spring Stallion Show. 1923.

Front page of a Peterborough Agricultural Society event that took place away from the main Showground. This event was advertised as being at the Repository which was close to the City end of Lincoln Road, Peterborough.

oldest engineering business in Peterborough. It certainly had a major bearing on the City's future development. As a regular exhibitor and competitor, it played a seminal role in developing the close and integral links between the commercial and political organisations of the City and its surrounding area, and the Society. It is the strength of these links which have always been very much a feature of the Society's activities, and which help to explain its success. The respect and affection with which James Barford was regarded was demonstrated by the fact that nearly 300 employees of the Queen Street Works attended his funeral.

The Show of 1923 saw the Prince of Wales, later to become King Edward VIII pay a visit in his capacity as President of the Peterborough Royal Foxhound Show Society. The local press heralded his visit with adulatory headlines such as these: 'The Best Sportsman in the Country at Peterborough'; 'City's Proud and Loyal Welcome of a Prince' and 'Thousands Throng the Streets to Greet the Future King'. The paper then went on to record:

> *A glorious day of perfect sunshine assisted to make the visit one that will long be memorable in the City's proud history. The Prince travelled on the ordinary service express from Kings Cross, and arrived in Peterborough at 11.37 am. An hour before people had assembled several deep on the pavements lining the route to the Agricultural show at Eastfield. In Broadway a banner stretched from Reed's Garage to the opposite side of the road, with 'Welcome To Our Prince' printed on it. As on previous Royal visits, the prominent members of the City Council and the Agricultural Society were formally introduced to His Royal Highness.*

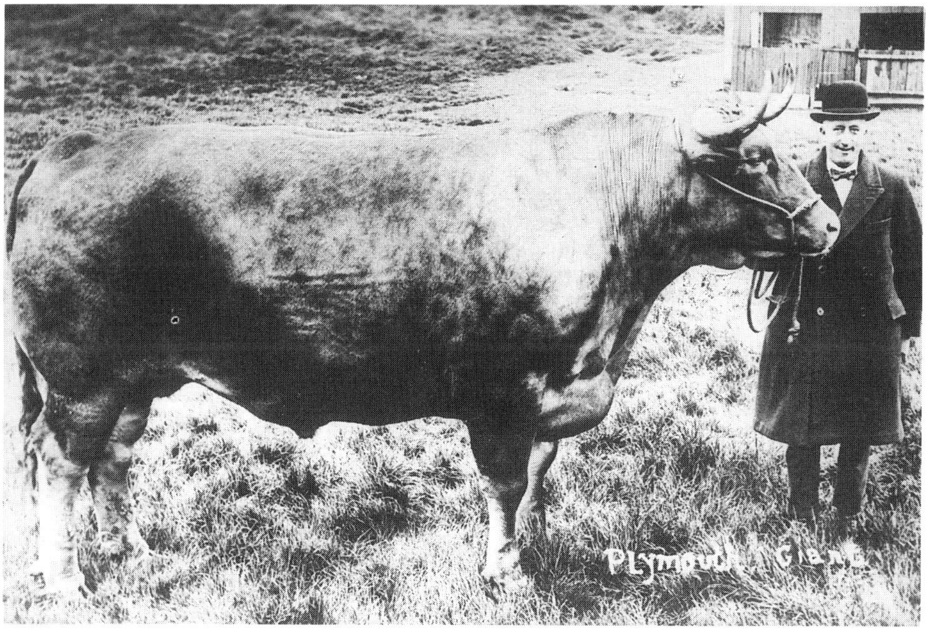

30. 'Plymouth Giant'.

This splendid animal weighing 32 cwt, was 17 hands high, 12 feet in girth, 15 feet long and 4 years old. It was a South Devon breed belonging to Mr Hill of Wakefield, Yorkshire who has certainly travelled a long way to exhibit his beast of which we can see he is very proud. Date unknown.

Right-hand side Main Avenue - showing Trade Stands having a busy time.

31. Eastfield Showground, 1933.

A picture along one of the main avenues at the 1933 Show at Eastfield with brisk business being done by some of the trade stands. This also gives the impression that some folks liked to promenade up and down purely for the pleasure of seeing and being seen!

Space does not permit a detailed survey of all the shows of the inter-war years. The 1930 Show however, was notable from the point of view of the judges as being the first occasion for many years that an objection was raised after an entry had won a prize. Captain Fitzroy, Speaker of the House of Commons, had two bulls, 'Foxhill Wild Piper' and 'Foxhill Lord Caryl', entered in the Yearling Pedigree Dairy Shorthorn Bull Class, 'Wild Piper' being placed first and 'Lord Caryl' second. A protest was then lodged that the latter had also won a prize at the Northamptonshire Show, a condition of the class rules being that no bull having taken a prize was eligible to compete for another Dairy Shorthorn Association prize again in the same year, except at the Royal Society's Show. Accordingly 'Lord Caryl' was disqualified, presumably much to Captain Fitzroy's chagrin, and Lord Dulverton's 'Batsford Wildeyes Treasurer' and 'Batsford Premier Foggathorpe III' moved up to second and third place respectively.

The major non-agricultural display in the Main Ring was a first for the Show, namely, the staging of a Military Tournament by the Inniskilling Dragoon Guards. They employed 78 N.C.O.s and men and about 50 horses. The particular animal that the crowd was looking out for, however, was the regiment's famous mascot, 'Fritz'. This was a grey pony mare 24 years of age, captured from the Germans during the regiment's charge at Harbonnieres on 18 August, 1918. During this charge, the regimental trumpeter of 'B' Company had his horse shot from under him but, with insouciant coolness, he then managed to commandeer this pony and to continue to gallop around, presumably trumpeting as and when required, for the rest of the battle. This Show was of an especially martial character since the band of the famous 2nd Battalion of the Gordon Highlanders were also on hand to provide the musical entertainment for the occasion. The editorial in the *Peterborough Standard* relating to the Show, carried a leading article on the historical background of the Society including the following statement:

In 1797, the Society's first show was held in the Waggon and Horses yard in Narrow Street.

32. The Eastfield Showground. 1936.

A general view of a part of the Showground, given over to the display of agricultural implements and it is of particular interest because one of the trade stands visible is occupied by the Caterpillar Company. This major international company recently became the parent company of the long-established Peterborough manufacturers of diesel engines now once again known as Perkins Engines Company Limited.

This is the first direct reference in a newspaper to the location of the Society's Shows in its earliest years. No evidence is actually put forward in support of this statement and in the absence of verifiable information it is best to keep an open mind about the precise location and the nature of the shows of the 1797-1803 period and indeed about any other shows that might have been held in the period before the re-inauguration of the Society in 1838 or until the historically verifiable use of the Waggon and Horses Yard in 1848.

On 3 July, 1931 the *Peterborough Standard*, reporting on the history of the Peterborough Agricultural Society's Showgrounds, made a similar, broader, but unsupported statement:

> *The 'Waggon and Horses' yard in Narrow Street provided ample accommodation for the first exhibition in 1797. But the rate of progress was amazing, even in those days. The next site was a bigger field, in Westgate, and after that the Old Show Ground at Millfield, now a housing estate, was taken. Twenty years ago a move was made to Eastfield, but even the broad acres of the original showground have failed to house all who desire to associate themselves as*

33. View across Eastfield Showground. 1939.

This view emphasises the scale and range of the agricultural implements and machinery exhibits at the Show. Because of its size, the Peterborough show attracted exhibitors and trade stands from right across the country and many nationally-known companies came every year to display their products.

> *exhibitors or tradesmen. So now another big strip of land has been acquired, and still the showground is by no means too big.*

This is the first intimation, at least so far as the press was concerned, that the Society was again faced with the possibility of running out of space and requiring once more to take on the daunting tasks involved in locating, surveying, negotiating, planning and developing new and larger accommodation.

At the 1931 Annual Luncheon, the Mayor, Councillor A. E. Fletcher, gave a speech which was very critical of the government's new proposals to create marketing boards for a number of agricultural products and the *Peterborough Advertiser* went on to record his views, as an admitted outsider, about the state of agriculture. He said:

> *'It stood to sense that when potatoes were fetching £1 a ton and cattle £12 to £14 per head, the farmer was not getting enough to cover the cost of production. As a consumer he hoped that the state of affairs would be adjusted so that everyone had a fair deal. It was not for the welfare of any community to force people into the bankruptcy court'.*

34. Group of Peterborough Agricultural Society officers. 1932.

Back: W.O. Snowden; G.W. Moore; W.S. Smith; M.R. Riddington; J.R. Horrell; E. Beck (Vice-Chairman); A. B. Gould; R. Bibby (Secretary); G. P. Little.
Front: J. N. Scorer; J. H. Makinson; W.Hemmant; F. M. Odam; I. Whitsed(Chairman); R.E. Dowse; A. J. Bird; W. R. Odam.

Agriculture in the 1930s

As with most other sectors of the economy, agriculture was hit very hard by the economic depression following the 'Great Crash' of 1929. Prices fell as surpluses built up, workers were laid off and both cereal and livestock farmers were affected, many going out of business while large areas of land became derelict. So serious was the situation that the Government was compelled to break with its commitment to non-intervention. Slowly and reluctantly, they introduced measures such as price support for wheat, a subsidy for fat cattle and the renewal of the sugar beet subsidy.

A radical departure from tradition was the imposition of tariffs and quotas on a variety of imported foods. These indirectly benefited agriculture while another interesting innovation was the introduction of Marketing Boards. After early trials, these Boards became effective after the passing of Agricultural Marketing Acts in the years 1931 and 1933. They allowed for the creation of what were basically producer-controlled cartels. The co-operation of the NFU with these

35. Non-Agricultural activities at an Eastfield Show in the 1930s.

This picture offers a flavour of the diverse activities that took place at the shows of the time. In the centre is a stand advertising the wares of Messrs Neaverson, pioneers in the Peterborough area of the sale of radios and phonographs. In the foreground a gentlemen is trying his luck with the post-horn, perhaps for the benefit of Mr Neaverson and his state-of-the-art audio equipment while the St John Ambulancemen stand by in case of emergencies.

was gained from the start and they applied to hops, potatoes, pigs and milk. They were welcomed at the time, and were generally considered to have been a success. We are left, however, with the feeling that Government intervention in agriculture was at best lukewarm and it was only the imminence of war in 1939 that convinced the politicians and mandarins that the slow response evident in 1914 could not be repeated and that every effort should be made to place the industry on an effective war footing, no matter what that might mean in terms of thoroughgoing state control.

The inter-war years had not been kind to Britain's traditional heavy manufacturing and extractive industries such as iron and steel-making, shipbuilding, engineering, textiles and coal mining or to the districts where these tended to be concentrated. Large areas of the North East of England, South Wales, Lancashire and Yorkshire and central Scotland, for example, were blighted by economic depression and industrial decline, environmental decay, mass

36. Beauty and the Beasts.

Date unknown but about 1930. The authors feel that all three shown in the picture have their good points.

37. 'It never rains but it pours'.

Braving the elements and presenting a brave face on things at the Eastfield Show on a bleak, windswept so-called summer's day. Exact date unknown.

unemployment and the human misery that accompanied these. On the other hand, considerable areas of the Midlands, East Anglia and the South East, continued to experience economic growth, based largely on the manufacture of consumer goods in light industry and the development of the service sector. Many who were employed in these parts of the country experienced a real increase in their living standards and prospects during this period. They became consumers able to enjoy a widening range of goods and services aimed precisely at people like themselves. Among the products they began to 'consume' in considerable quantities was the countryside itself. This product was becoming increasingly accessible because of suburban and rural housing development and the steady rise in car ownership.

A veritable industry began to emerge as the urban middle classes became more and more attracted towards the pleasures which they perceived as being offered by country life. Guidebooks and somewhat sentimentalised accounts of rural life poured off the presses, to be avidly snapped up by town-dwellers with deep pockets and an eye for the picturesque and quaint. An image was created of an idyllic England. Medieval church spires embowered in well-matured oaks

38. Royal Visitors to Eastfield. (1)

Left to right: The Duke of York, later King George VI; Mr W. Hemmant, Chief Steward; Mr I. Whitsed, Chairman; The Duchess of York; Edmund Beck, Vice-Chairman.

sheltering colonies of raucous rooks and enclosed in churchyards full of venerable, gnarled yews. Village idiots supping their pints of mild-and-bitter on the weather-beaten benches in front of an ancient timbered, thatched pub overlooking the green where the village's young bloods and past masters are engaged in an epic no-quarter-given-nor-expected cricket match with the old enemies from the neighbouring parish. Close by are the maypole, that ancient symbol of fertility, a duckpond and knots of innocent children making daisy chains. Needless to say, this scene of Arcadian delight was given the final tincture by the compulsory sun and deep blue sky. And so it went on.

Clearly such a pastiche, while containing elements that were recognisable and realistic, had very little to do with how those who lived in the countryside went about earning an everyday living. Nevertheless, this is how many town-dwellers came to see the countryside and its villages and settlements. Not only did they often drive through it or walk over it, seeking recreation, but increasingly they came to live in it, seeking that elusive thing perceived as a better quality of life than that which was seen as available in the towns. In doing so they contributed

39. Royal Visitors to Eastfield. (2)

The Duke and Duchess of York show their interest in the cattle classes at the same Show as illustration 38.

40. Shoeing Competition

Farriers display their talents in the Shoeing Competition at the Eastfield Showground. For some reason, competitions in this highly skilled craft rarely reached the limelight and few photographs of these events were published in the press.

unwittingly to the destruction of the very idyll they were so eagerly seeking. Their obsession with all that was picturesque and precious in rural England contrasted starkly with their general lack of interest in and knowledge of agriculture and the rural way of life. It is certainly true that Agricultural Societies played a major role in educating the general public about farming and country ways but it may well be that it took the dire circumstances of war to bring home the fact that the aesthetic and historical attractions of the countryside were most definitely subordinate to its prime purpose, in war or peace, which was the production of food for Britain's large and overwhelmingly urban population.

The Society's Shows before the Second World War

In 1935 a large and sophisticated piece of machinery drew much attention at the Show. This was the 'Gyrotiller' exhibited by Mr N. Box of Peterborough and was truly a machine designed for all seasons. It weighed 15.5 tons, had a working pressure of 12lbs per square inch and was said to be capable of ploughing, dragging and harrowing simultaneously. Mr Box modestly explained that he had six of these contraptions available and that they could be hired from him over an area extending from Newark and Skegness down to St Neots.

41. 'Fashion on Parade, Peterborough style.'

While some people went every year to Royal Ascot , some to Henley and others to Wimbledon, at the Peterborough Show, Wednesday was Ladies' Day and the time and place for the fashion-conscious to parade up and down, showing off their latest finery. Exact date unknown.

It was evidence of the troubled economy that the National Employment Exchange Service for the first time had a stand at the Show. Many of the temporary staff employed by the caterers and in a variety of other roles by the Society at the Show, had been supplied by the local employment exchange. One of the reasons for exhibiting at the Show was to try to get away from the

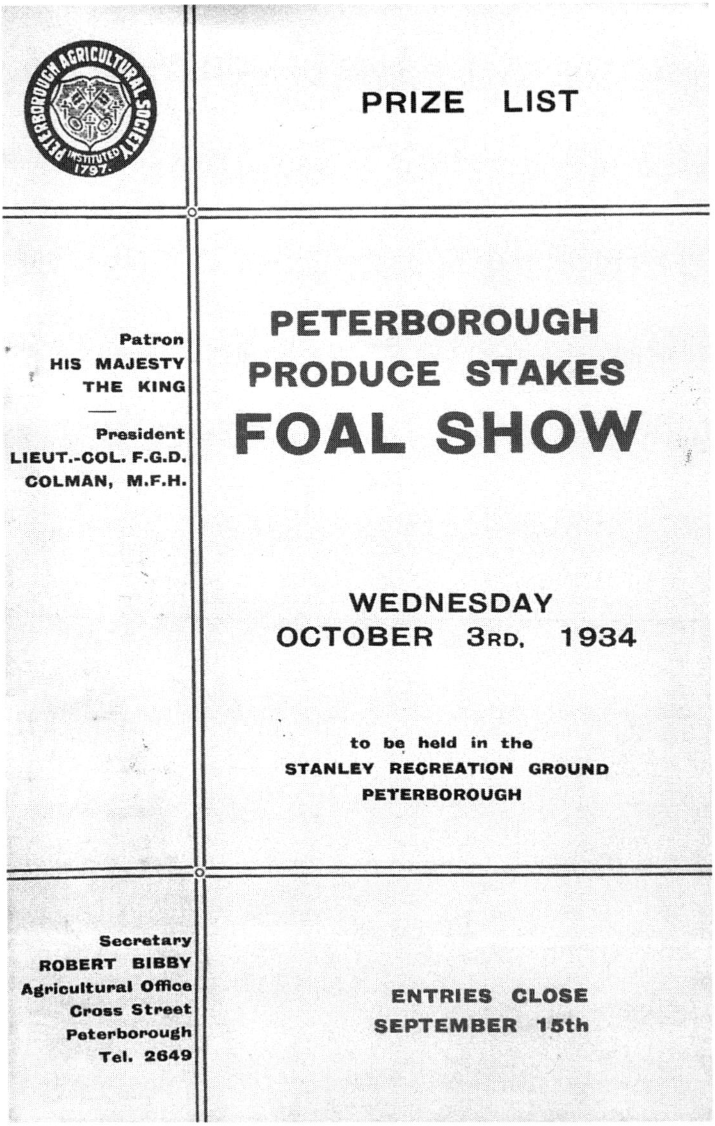

PETERBOROUGH AGRICULTURAL SOCIETY INSTITUTED 1797

PRIZE LIST

Patron
HIS MAJESTY
THE KING

President
LIEUT.-COL. F.G.D.
COLMAN, M.F.H.

PETERBOROUGH
PRODUCE STAKES
FOAL SHOW

WEDNESDAY
OCTOBER 3RD. 1934

to be held in the
STANLEY RECREATION GROUND
PETERBOROUGH

Secretary
ROBERT BIBBY
Agricultural Office
Cross Street
Peterborough
Tel. 2649

ENTRIES CLOSE
SEPTEMBER 15th

42. The Peterborough Agricultural Society's Produce Foal Stakes Show, 1934.

The picture shows the cover of the prize list for the above event. Notice that this Show was held some while after the main Annual Show and was staged at the Stanley Recreation Ground.

43. First Day at the 1935 Show, Eastfield.

This picture captures the judging taking place in Class 45 for Percheron Mares in foal or with foal at foot.

impression that these Exchanges were just places where the unemployed could pick up their dole money and to show that they could actually help employers in the farming community to recruit new staff as well as to assist would-be workers to secure employment.

The *Peterborough Standard's* assessment made after the 1935 Show recorded some statistics compiled by the Society which showed a steady rise in admissions through the turnstiles and in gate receipts. These increased from £1,278 1s 0d in 1933 to £1,418 6s 2d in 1934 and £1,592 19s 6d in 1935 while the number of members of the public attending rose from 13,474 in 1930 to 20,567 in 1935.

The 1939 Show saw the ceremonial inauguration of the 'The Beck Memorial Bandstand' which was the Society's very tangible tribute to the late Mr Edmund Beck. He was a well-loved and flamboyant character who had served the Society as Vice-Chairman and in many other capacities. A challenge cup was also generously presented by Mr K.G. Fryer, a future Chairman of the Society who had long been Mr Beck's partner in the eminent local firm of Fox and Vergette, surveyors and estate agents. Among the titbits of information in the *Peterborough Standard* concerning this show was that re-paved pathways to the main entrance

44. Judging the best Stallion Hounds. 1938.

45. Broadcasting from Eastfield.

An early live broadcast from the Eastfield Showground on the BBC Midland Regional Service. Edmund Beck (Vice-Chairman) in the centre accompanied on his right by the Chairman of Council, Alderman I. Whitsed and left, Mr J. R. Major, Council Member. 1938.

46. 'The Madding Crowd'.

A view of crowds milling around at the Eastfield Showground: exact date unknown but during the 1930s. The Annual show has always been the place to catch up with old acquaintances including some that the members, exhibitors and visitors probably haven't seen since the previous year's Show!

in Eastfield Road met with general approval but most of all from the ladies because *it averted discomfort to the feet and clouds of dust to soil dainty dresses.* The Ministry of Agriculture had an impressively large stand containing a variety of hints about ways to market field and garden produce. This is perhaps an indication of the widespread feeling that war was imminent and that the country might again be subjected to a blockade. In that sense, therefore, it anticipated the 'Dig for Victory' Campaign.

The 1939 Show was widely regarded as a very successful one, the paying attendance being no less than 14,549 and the records show that the caterers, Messrs William Ely and Sons, served over 1,700 luncheons in the first and second class dining tents during the three days of the Show.

An interesting insight into the service provided by the Post Office during the Show is recorded in the following report from the *Peterborough Standard:*

> *The travelling Post Office has not been provided this year but similar services have been provided which have proved to be most adequate. There have been three deliveries of mail a day, including two for parcels, and the office has been open from 9 a.m. to 7 p.m. covering practically*

all the normal Post Office services provided by the Peterborough Head Post Office. The volume of telegraph work was rather greater than last year. For instance over 100 telegrams were sent and over 6,000 words of press messages. There were three telephone kiosks and the office was at the main gate.

Despite the obvious success of the 1939 Show, the people of Peterborough were left with the tension and uncertainty that was hovering over the entire nation about the outcome of the British Government's appeasement policy towards Hitler. They were allowed no further doubt, however, after the Prime Minister, Neville Chamberlain made his portentous radio broadcast on the morning of the 3 September, 1939 announcing a state of war. It was not to be until 1946 that the Society would once again stage an Annual Show.

47. Eager Anticipation.

A scene showing packed crowds evidently full of interested anticipation as they wait patiently for an event in the Main Ring at the Eastfield Showground. Both men and women sport a variety of headgear and it seems to have been de rigeur for the men to wear collars and ties. A deterioration in standards of dress and pride in personal appearance would be all too obvious if this picture was to be compared with a present-day scene at the Show.

48. Souvenir brochure for the 1937 Show.

For an all-too brief period during the 1930s, the Society produced colourful landscape format souvenir brochures for the Annual Show which now attract the interest of collectors of ephemera.

49. Presentation of Prizes. 1939.

H.R.H. The Duchess of Gloucester presents the Champion Gold Cup and the Maxwell Challenge Cup to Mr J. R. Hindley, the owner of 'Bradley', the Champion Riding Hunter of the Show.

50. The Edmund Beck Bandstand.

Crowds cluster around the Edmund Beck Bandstand at the Eastfield Showground while the band takes a break. This distinctive bandstand erected in recognition of the great contribution made to the Society's affairs by Edmund Beck was paid for by Mr K.G. Fryer, his business partner. It was later dismantled and then rebuilt on the Alwalton Showground.

51. 'Judging the Great Whites.'

Here we see the judge, Mr E. Tomlinson, casting an final expert eye over the entrants in one of the porcine competitions. 1938.

52. Display of Agricultural Implements.

A pictorial view of the Eastfield Showground giving some idea of the extent of the agricultural implements on display.

53. The Splendour of the Heavy Horse

'Decoy Sunshine', Gold medal Winner at Eastfield, 1939.

54. The Eastfield Showground from the Air.

An aerial view of the Society's Showground at Eastfield, date unrecorded but probably just before the Second World War. This view is taken looking in a westerly direction with Eastfield Road in the foreground. Notice how even at this stage the motor car is becoming a major planning consideration for the organisers and it is easy to understand that it was the shortage of suitable space for parking that hastened the demise of the Society's Shows on this site. Most of this land has subsequently been built on and includes, centre left, the site of what is now the Peterborough Regional College.

The World War 2 Years

The Society's activity did not cease altogether during hostilities. On 25 July, 1940, the local paper ran a headline *'Farmers' Red Cross Sales'* and went on to describe a sale of live and dead farming stock which included: 20 cattle; 40 sheep; 30 pigs and 100 poultry and geese in addition to various devices including binders, cultivators, harrows, a weighing machine and weights, sundry implements and tools and a pony float and harness. The sale took place at Milton, the family estate of the Fitzwilliams. That this was not purely a business affair and that some levity was being allowed in these otherwise dark times, is indicated by the sale being accompanied by a fete, side shows, refreshments and licensed bar. Musical entertainment was provided by the local Castor and Ailsworth Silver Band. Apparently large crowds attended and enjoyed themselves but this was the nearest they would get to an organised farming show until the month of July 1946.

PETERBOROUGH
PRODUCE STAKES
FOAL SHOW

WEDNESDAY
OCTOBER 2nd, 1940

Entries Close Wednesday, September 18th

The Show will be held in a Paddock at the corner of Craig Street and Lincoln Road, Peterborough, about 100 yards from the Repository

Secretary:
ROBERT BIBBY, Agricultural Office, Priestgate, Peterborough
Telephone 2649

55. Catalogue for Peterborough Produce Stakes Foal Show, 1940.

Documentary evidence of the Society's activity on yet another site, in this case on land at the junction of Craig Street and Lincoln Road close to the site of the old Boroughbury Showground. Note that this event took place during the Second World War when, for obvious reasons, the Society's activities were considerably curtailed.

As in the First, so in the Second World War, agriculture underwent a transformation but this time there was a difference. The State intervention in the economic life of the country which was a feature of wartime conditions was to continue after hostilities had ceased and in a form which was intended to guarantee that there would be no return to the long-standing depression from which agriculture had been suffering. Laissez-faire was finally expunged and the State decisively took on an active responsibility for creating the conditions in which the nation's farmers could enjoy a far greater degree of stability and prosperity. Farming in Britain was henceforth to find itself increasingly under the direction of Parliament, of Whitehall and later, towards the end of the century, more controversially, of Brussels. Many saw this as a mixed blessing.

Submarine tactics and technology had progressed markedly and these vessels constituted a much more serious menace to the United Kingdom from 1939 than they had in the previous war. The destruction that they were able to cause among Allied shipping created an appalling threat to Britain with her dependence on imported food supplies. It was considered imperative to bring about a vast expansion of home output. So the Government applied a mixture of the stick and the carrot to Britain's farmers. Guaranteed prices were a great incentive for many while those farmers unwilling to comply found their autonomy gone as they came under the jurisdiction of local representative committees, in the members of which, it has to be said, the local farming community did not always have total faith.

The war years witnessed far-reaching changes in the agricultural industry. Sizeable tracts of neglected and under-used land were reclaimed and a record acreage of marginal and unused land was brought under cultivation. Large quantities of livestock were slaughtered to enable savings to be made on feedstuffs. Compulsory cropping orders were made and farmers forced to mechanise. While this may all sound very draconian, the guaranteed prices were actually generous and the farmers prospered as a result. Even the farm workers, benefiting as they did from their temporary scarcity value, saw their real wages rise to unprecedented levels. In this situation it is not surprising that both sides of the industry readily accepted both the necessity for and desirability of, some state control. In pragmatic terms, the Government's intervention can retrospectively be judged as a successful one. Output expanded sufficiently to enable the country's population to be fed and even to achieve a high standard of nutrition, although the diet was widely considered to be monotonous.

There is no doubt that the public perception of the farming industry was that it had done an enormous amount to assist the war effort. Its strategic importance for the nation's survival had been evident for everyone to see and there was a widespread feeling that the country's drift towards an increasingly unfavourable position in world trade would at least be checked if greater attempts were made to achieve self-sufficiency in staple food requirements. It is therefore hardly surprising that the Agriculture Act of 1947 met with a general welcome for it maintained the incentives embodied in guaranteed prices while requiring farmers to employ good practice. It was also popular with the consumer because it encouraged cheaper prices for many of the products of British agriculture.

The Return to Peacetime Conditions

Clear and welcome evidence of a return to normality was recorded on 5 July 1946 when the *Peterborough Standard* gave notice of the forthcoming Annual Show under the following headlines:

> *Two-day Victory Summer Show next week - more than 900 stock entries; hounds from 36 packs.*

It continued:

> *With entries well up to those of the Nineteen Thirties, the Peterborough Agricultural Society's Show at Eastfield next Wednesday and Thursday promises to rank with the best seen in this country since the War. The six years break inflicted by events has not caused interest or enthusiasm to lapse or even lag. Everything in fact, points the other way - to a happy repetition of what happened after the 1914-18 War when many former records were surpassed and the attendances topped 20,000 for the three days.*

This was not mere hyperbole because the Show did indeed prove to be a memorable one from the point of view both of the Society and of the citizens of Peterborough. Although necessarily conducted under the constraints of austerity, it was quickly shown that the Show had lost none of its appeal as a public attraction and well over 7,000 people flocked to the Eastfield Showground on the Wednesday which was the opening day. One notable double award win in the Draught Horse competitions was achieved by Mr J. Humphrey, of Gorefield, Wisbech. His bay filly, 'Bleak House Iris', won both the open two-year-old filly or gelding class and also the class for the best animal within 30 miles of Peterborough.

In 1947, the Society's Show gave every appearance of matching or even exceeding its pre-war best. The attendance on the first day was a record-breaking 8,910 and many of these were able to enjoy the Royal presence at the Show. The very large crowds who thronged Bridge Street, Long Causeway, Broadway and the Market Place in order to see the Duke and Duchess en route, were frustrated when the Peterborough police diverted the Royal cavalcade around the almost empty side streets on its way to the Eastfield Showground. The grandiose scale of the 1947 Show entry classes is indicated by an estimate that the final parade of winning animals contained beasts worth a total of nearly one million pounds.

After the war, the policies of successive governments towards agriculture were intended to create conditions favourable for the production of more and cheaper food. The outcome was a remarkable increase in farm productivity. The State provided circumstances in which farmers were encouraged to increase their output, productivity and efficiency, in the knowledge that they had the security which came from price stability. The effects of this proved to be far-reaching. The horse virtually disappeared from the world of husbandry while the ranks of the farm labourers were also decimated, both being pushed aside by the urgent drive to mechanise. Selective plant and animal breeding produced hitherto unprecedented increases in output. Applied science could provide farmers with

56. Characters at the 150th Anniversary Show.

Stalwarts of the Society taking a few minutes off to relax at the 1947 Show which was celebrated as marking the 150th Anniversary of the founding of the Peterborough Agricultural Society. Left to right: Mr A.B. Gould, Senior Steward; Mr R.V. Cobb and Lt.Col Part, MFH., Judges.

more effective animal feeds and fertilisers while animal care was rendered more efficient by increasing use of pesticides and vaccines. The marked tendency which had already been seen elsewhere in business towards larger-scale, ever more capital-intensive and specialised production was replicated in agriculture with the result that the very ugly new word ‘agribusiness’ made its appearance, appropriate to the current scale and nature of much of the industry. In fact substantial numbers of farmers found that they were totally in thrall to large food-processing companies, many of them multinationals, their purchasing power being such that they could easily dictate not only what was produced but the prices that producers would receive for their efforts. These developments were not without their critics because they had very profound effects on the nature of life and work in the countryside as well as on the actual appearance of the countryside itself, much of it now bereft of hedges and the bulk of its wildlife. While British agriculture became both highly cost-efficient and prosperous, traditional mixed farming declined and the advantages, as expected, tended to accrue most of all to those farmers or organisations who were large enough to benefit from economies of scale.

57. More Peterborough Agricultural Society characters.

Left, Mr H. Le P. Grimwade, Vice-Chairman of Council and Mr K.G. Fryer, Member of Council and future chairman.

While their numbers were being decimated, those farm workers who remained on the land were being called upon to learn and apply a whole range of new skills required because of the extensive spread of new technology. This they did most effectively but without seeing much benefit in terms of increased wages, status or security. In fact, between 1946 and 1976, the number of full-time agricultural workers fell from nearly 750,000 to less than 200,000. The farm workers' dream of parity with urban industrial workers remained but an aspiration and the changing nature of the work they had to do all too often meant replacing valued camaraderie by enforced and unwelcome isolation.

In the Peterborough area in the post-war period, the diesel engine manufacturers, Perkins, paid wages which were high enough to lure labour away from the land much to the consternation of local employers of agricultural workers. The first record of combine harvesters being used in the Peterborough area was in 1946 and it was not until well into the 1950s that they came into general use along with other forms of labour-saving machinery. The fact that so many workers had already left farming, somewhat softened the blow as far as later redundancies were concerned.

During the 20 years after 1945, the English village as an occupational

58. Royal Visitors, 1948.

The presence of Royal Visitors always draws crowds to the Show. Here we see Their Royal Highnesses the Duke and Duchess of Gloucester gazing approvingly at a miniature bus powered by a 2-stroke engine which was used to convey children round the Alwalton Showground.

community almost disappeared. It found itself under a twin-pronged attack from encroaching urbanisation and the drift away from the land of the rural labour force. The social changes of the 1930s which had brought townspeople to visit the country or to settle there in search of some rural idyll accelerated, a major agency of this movement being the motor car. Now they came in droves to reside in the villages, usually on estates of owner-occupier houses which failed to blend in with their surroundings. These villages were little more than dormitories for middle-class workers whose employment was to be found in towns and cities that were within reasonable commuting distance. The new country dwellers often still had urban roots and loyalties and the money and the means of transport which allowed them to maintain their roots outside the village. They often made relatively little use of local shops, post offices and bus services and the indigenous population of the villages, many of whom were poor and maybe old as well, had to face a declining infrastructure of local services as well as finding that their own children could not afford to buy houses locally because increasing demand was pushing prices out of their reach. The serious economic, social and

59. 'Anwick Benjamin.'

The winner of the F.W. Griffin Cup in 1948 for the best Lincoln Red Bull at the Show. Despite the obvious pride that he must be feeling by being able to exhibit such a splendid beast, does the owner's stance not suggest some degree of trepidation? 'Anwick Benjamin' for his part looks slightly bored by the proceedings.

cultural implications of these developments were making themselves felt in the countryside of the Peterborough region during what was a period of success and relative prosperity for the farmers of the area. They form the background to a situation in which the Peterborough Agricultural Society found itself in the enviable position of recording steadily increasing numbers of exhibitors, competitors and paying visitors at the gates.

In 1950, the BBC used a mobile recording studio for the first time at the Annual Show. On the Wednesday they brought a number of the Society's well-known personalities to the microphone and in the Home Service programme 'Country Lover', broadcast the next day, Godfrey Baseley presented 'A Picture in Words' of the Show. It was obvious that he was very impressed by the Show and its efficient management. After the recording had finished, he made a tour of the Show and paid a visit to the stand belonging to Messrs Hayward and Towell of Peterborough where their slick sales patter persuaded him to splash out and buy an electric 'Wolf Universal Tool'.

Outstanding features of this particular Show were the magnificent entry of riding hunters and a very strong showing of Ayrshire cattle, the growing popularity of the latter at these events meaning that they were running Dairy Shorthorns a close second. The entries for cattle and pigs had never been surpassed. Five men who between them had 266 years on the land were honoured to receive long service awards from the Royal Agricultural Society of England which were presented to them by the Marchioness of Exeter. They included two brothers, Samuel and George Pridmore who had been employed for 58 and 56 years respectively by the Proby Family at Elton and Samuel Trayford who had no less than 61 years service on a farm at Blatherwycke. A new feature at the 1950 Show were two sixty-foot TV masts erected by the Electrical Contractors Association which exhibited many different television models. These were still very much at the stage of being a novelty and crowds gathered round goggle-eyed, in some cases seeing a television for the first time as they watched the programmes which at that time were transmitted from Sutton Coldfield. One of the local TV dealers on this stand was Mr A. Stokes of the Old Arcade in Westgate, Peterborough and so successful was he that by lunchtime on the first day, he had apparently sold three sets priced at a total of over £200 and had therefore already paid the rental for his display!

60. A Society Horse Show.

Exact details relating to this photograph are not known but the scene appears to be the late 1940s and shows the Stanley Recreation Ground. Research has so far proved it impossible to determine whether the Society held its Horse Show on this site because of problems at the Eastfield Showground and whether the holding of this event was a regular annual occurrence. Clearly this is a winter scene.

For the first time since it came into being in 1933, the firm of F. Perkins Ltd displayed its Peterborough-made diesel engines at the Show. They were having a busy time that year and had already appeared at four other Agricultural Shows. Their exhibits, which had been judged the best of their kind at the prestigious Bath and West Show, included three chromium-plated diesel show engines, one of which was sectionalised to demonstrate its cycle of operation. It was turned by an electric motor.

The 1951 Show was notable because of the visit of HRH Princess Elizabeth, the heir to the throne who was soon and unexpectedly to find herself on the throne. She had spent the previous night with The Duke and Duchess of Gloucester at nearby Barnwell Manor, their family residence and she subsequently travelled into Peterborough in a motorcade along the Oundle Road. Large crowds turned out to cheer her on her way into the City and through it to the Eastfield Showground. As the *Peterborough Standard* reported:

> *...they passed through lines of people who with loyal fervour, waved flags, fluttered handkerchiefs or doffed hats as the Princess passed by. Her Royal Highness acknowledged the salutations with smiles and waves of a white-gloved hand. Standing almost alone on Botolph Bridge, near the city boundary, a small boy dipped a huge Union Jack in dignified salute.*

61. Fordson Major Tractor.

This typical tractor of the early 1950s is a Fordson fitted with a Perkins L4 Conversion engine. Note the well-known Perkins Engines radiator badge/logo.

Further along Oundle Road, near Bread Street, about 160 children from Orchard Street Secondary School who had expressed a wish to see Princess Elizabeth cheered loudly as the Royal visitor drove by. The reporter continued, *Travelling as we were behind Princess Elizabeth's car, we were afforded ample evidence of the reception given to Her Royal Highness. Every few seconds we heard such exclamations as , 'Isn't she pretty?' 'Isn't she lovely?' And so she was.*

The Champion of England Gold Challenge Cup awarded for the best hunter in the riding classes was regarded at the time as the Show's premier prize and this year it was carried off by the formidable 'Mighty Atom', a seven-year-old chestnut gelding which had been runner-up in this class at last year's Show but had gone on to surpass itself by being unbeaten at fourteen subsequent local and county Shows. Among the notable personalities on hand at the 1951 Show was Mr 'Billy' Geddes, who, at the age of 85, could be described as 'The Grand Old Man of the Peterborough Show'. He was the local agent for the firm of Chambers and Fargus and on the Tuesday he opened the company's stall for the 43rd time at an Annual Show. He had attended his first Peterborough Show in 1899 and proudly boasted how he had never missed a Show since that year. He commented on how the Peterborough Show had expanded and improved and gave much of the credit for this to Robert Bibby who was then approaching the end of his long and successful service as Secretary to the Society.

62. The Royal Visitor, 1951.

H.R.H. Princess Elizabeth, soon unexpectedly to find herself on the throne, with Lord and Lady de Ramsey and Robert Bibby, the Society Secretary, about to present the major challenge cups.

63. Presentation of the Champions' Cups, 1954.

This picture shows the second day of the 1954 Show at Eastfield and captures the major prize-winners who are about to receive their awards. On the left is W.S. Smith, Chairman of the Society's Council with Lady Ethel Wickham, President in that year and still going strong at the age of 90. In the centre is Robert Bibby, Society Secretary.

In 1957, Robert was tragically to suffer a cerebral haemorrhage, dying without recovering consciousness. More will be said about him in Chapter 9 but his enthusiasm and drive had seen the Society enormously enhance its standing in the region and develop its Show from a two-day into a three-day event. More than an era passed with his death and the Society showed its gratitude to him the following year when the 'Robert Bibby Memorial Clock' was erected on the eye-catching gateway located at the main entrance to the ground in Eastfield Road. Incidentally, Robert Bibby was the last Secretary of the Society who in addition acted as Secretary of the local National Farmers' Union Branch.

The 1957 Show then, was the first to be organised and presented under the stewardship of the new Secretary, Roy Bird. He threw himself into his role with great enthusiasm and was rewarded by a record entry in the livestock classes. The weather adversely affected attendance that year. It rained so much that the car park at the Showground was an absolute quagmire and visitors were advised to park in the streets around the area. The local constabulary gave special

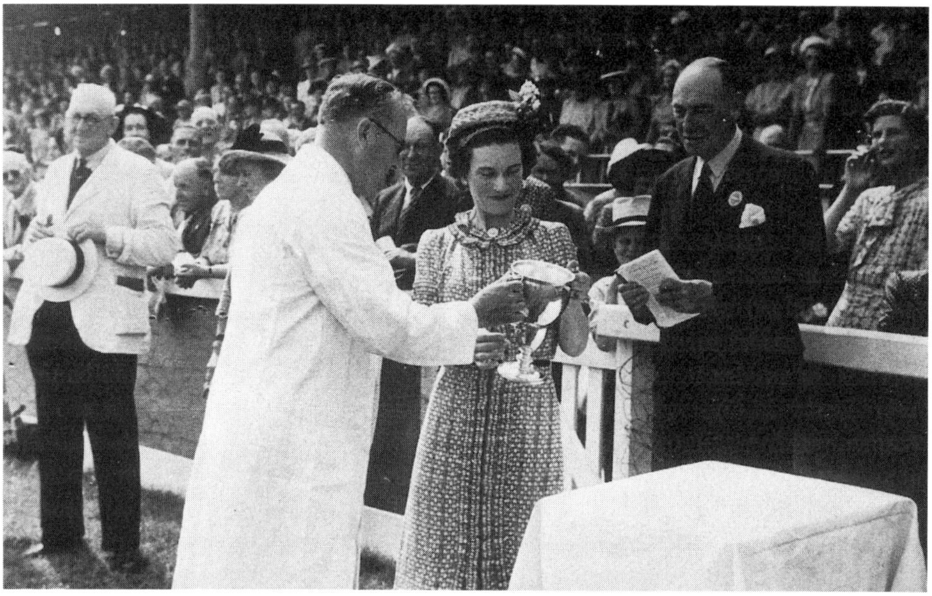

64. Prize-giving at Eastfield.

HRH Princess Alice presenting prizes with Mr. W. Hemmant, Chief Steward to her left while Society Secretary Robert Bibby stands nearby clutching a natty line in headgear. Date unknown.

permission for vehicles to be parked along the grass verges of the Dogsthorpe Housing Estate which was close by. A bulldozer was hired to try to make the surface of the car park flat but even this machine fought a losing battle. The world famous showjumper, Alan Oliver, riding 'Red Admiral' won a dramatic jump-off in the final round of the Peterborough Grand Championship to retain the 200 guineas Cobb Cup and secure a first prize of £100.

Both the following Shows, 1958 and 1959, suffered severely from inclement weather which caused severe damage to the Show car parks, which were at other times used as school playing fields. It was after the 1959 Show that the Vice-Chairman, Mr K. G. Fryer, suggested that the Society should look for another site which would provide sufficient land for the Show itself and adequate car parking, not only for the present but for the foreseeable future. The Society's long-standing, respected and influential Chairman, W. S. Smith, OBE, was not happy with this suggestion but made it clear that he would not stand in the way of progress and negotiations around the purchase of a new site at Alwalton took place during the period from 1959 to 1961.

Mr. Smith's stance was especially laudable given his devotion to the Eastfield Showground. Much of the labour force which prepared the ground every year for the annual show was drawn from workers on his farms and he personally defrayed much of the cost, the Society benefiting considerably from his generosity.

65. 'Up and Over.'

A picture exemplifying perfect co-operation and co-ordination between man and beast. The show-jumping events at the Society's Annual Shows always attract a great deal of interest and often feature top class performers.

It was also Mr. Smith who, while Chairman, developed the system of 'Assistant Stewards'. In theory these were newcomers prepared to learn the hard way and to look forward to the day when they would be elevated to the dizzy heights of Steward. In practice, however, these volunteers might spend as many as fifteen years winning their spurs. It also says much for his tireless work on behalf of the Society that during his tenure of office as Chairman, trade stand applications rose so rapidly that many had to be put on a waiting list.

It is also interesting to note that his Vice-Chairman, Mr. Fryer, was later to persuade Council to issue an instruction that all Stewards should wear grey hard hats at the Show. Some of those who felt this would give them an appropriate air of distinction were rather surprised, not to say, uneasy, when they discovered how much these items of headgear actually cost.

In 1960 a Royal Visitor for the first time arrived at the Show by helicopter. This was The Duchess of Gloucester who should have been accompanied by The Duke but he, unfortunately, was suffering from a slipped disc. All previous records for

66. Poster Art - 1960s style.

Illustration showing a splendid poster advertising the 1961 show and now an historical document in its own right.

67. Poster for 1963 Show.

A visually less attractive but no less historical poster produced for the 1963 Show designed when the sands of time were definitely running out for the Eastfield Showground.

attendance and entries were yet again broken at this Show, a popular feature of which was the spectacle put on by the RAF's Ceremonial Drill Team. It performed a continuity drill display in which well over 140 movements were carried out with precision and without a single word of command being uttered.

Some senior members of the Society recall various amusing events during the last years of the Eastfield Showground. On one occasion there was a sudden and

68. Peterborough Agricultural Society Council Members, 1964.

Back row: J.A. Wright; S. Harris; R.W. Bird (Secretary); G.P.L. Little; L.B. Odam; A. Hemmant; W.R. Riddington.
Centre: M. Smith; J.G. Hunter; A.B. Wilkinson; H.M.T. Jones; J. Scorer Jnr; J. Hemmant; M.W. Jeffreys; S. H. Gould; F.C. Mitchell (Treasurer); J.R. Horrell; C.E.W. Saunders.
Front row: H.C. Horrell; M.R. Riddington; G.R. Makinson; K.G. Fryer (Chairman);W.S. Smith, O.B.E. (President); J.Odam (Vice-Chairman); J.M. Odam; J.N. Scorer; A.B. Gould.

alarming clatter from the tent containing the canine events when a number of particularly large dogs, variously described as Irish Wolfhounds or Deerhounds, presumably became bored or irritated by being forced to wait while judging took place. They took it into their heads to make a sudden bid for freedom. Still attached to the benches on which they were being judged, they charged out of the tent and across the densely packed Showground, causing mayhem as the benches crashed into every conceivable obstacle, living and inanimate. The dogs were eventually subdued fortunately without any serious injuries, human or canine. On another occasion a spectacular parachute display went badly wrong when one woebegone parachutist, intending to drop skilfully onto a marked spot in front of the admiring crowds, instead passed over the Showground and out of sight only to come to earth not far away on one of the tennis courts which were situated at the front of what is now Peterborough Regional College. Unfortunately for him, the tennis court was locked. Its high fence made it impossible for him to climb out and it was some time before anybody discovered he was there and was able to rescue him.

1965 saw the last Show to be held at Eastfield, the decision having been taken some years earlier to move to a greenfield site at Alwalton with the intention that the Society should buy the land there and have sufficient space to avoid the necessity for yet another relocation in the future. The Society's success meant that it had outgrown the Eastfield ground as indeed it had outgrown its previous sites and history repeated itself as the growth of the City meant that housing and other urban development limited the potential for expansion at Eastfield. Not the least of the problems at Eastfield was the shortage of car-parking space sufficient for a Britain in which car-ownership was becoming the norm and also the fact that the car-parking space there was not the Society's own.

A Site Development Committee was established composed of the following members: Mr Joe Odam (Chairman); Messrs K. Ellis (Architect); K.G. Fryer; A.B. Gould; S.H. Gould; S. Harris; J.R. Horrell; M.W. Jeffreys; H.M.T. Jones; G.P.L. Little; F.C. Mitchell (Treasurer); W.R. Riddington; C.E.W. Saunders; J. Scorer; J.N. Scorer; C.D. Tebbs; H. Waterworth.

The Society was offered the choice of having the Alwalton Showground left outside or included within the designated area for Greater Peterborough. After some deliberation it opted to have it included within the area. This decision proved to be an extremely wise one because the Peterborough Development Corporation was therefore able to include within its road and parkway plans, adequate provision for the traffic that was likely to be generated by the various activities held at the Showground. Comparatively easy road access to and from the new Showground has undoubtedly been a factor contributing to the ongoing success of the Show's activities based at Alwalton.

The headlines recorded, 'The End of Show Pageantry at Eastfield', and 'A Squelching Climax to over 150 Years of Local History'. The article goes on:

The pageant of Peterborough's history since the Agricultural Show began in 1797 slithered to a halt in the mud-soaked main arena at 9.25 on Wednesday night. Rain was falling, most of the crowds had gone - and the spectacular navy finale, in which 1000 people were to have taken part, was yet to come.

The great quarter-size replica of the Wickstead Galleon, Flags Flying, Lights Burning, Crew Waving, stuck helplessly in the mire and dozens of scouts and cadets helped to push it off, as the producer and commentator, Mr Barry Simons, said sadly: 'We do apologise again, ladies and gentlemen, for the fact that we could not display to you our great finale, but there it is - the weather has beaten us'.

The Showground certainly resembled a quagmire. Several parts of the ground were literally knee-deep in mud and over 200 tons of straw were used, being scattered to a depth of 12 inches in the worst affected areas. Those who enjoyed the socialising side of the Show, taking a drink or two with old acquaintances, were forced to stand on chairs.

The Royal visitors on this occasion were once again the Duke and Duchess of Gloucester and it seemed fitting that the President of the Agricultural Show on this last usage of the Eastfield Showground should be the 10th Earl Fitzwilliam, the direct descendant of the 4th Earl Fitzwilliam who was the founder and first President of the Society when it was originally established in 1797.

Among the stirring sights at the 1965 Show were the superb displays of horsemanship put on in the main ring by the Royal Danish Hussars. They were appearing for the very first time outside Denmark where they acted as the personal bodyguard to King Frederick. A new Challenge Cup provided by the Duke of Gloucester for the best entry of the Guernsey Breed of cattle was won by Horrell's Farmers Ltd, a long-standing local farming family who had close associations with the Society and its Shows. The medals awarded for new implements, improvements or inventions applicable to agriculture included some for local firms. These were: Peterborough Motors, Tractor Division, who displayed four Ford tractors; Root Harvesters Ltd, a duplex potato harvester and F.A. Standen and Sons Ltd., of Whittlesey, exhibiting a two-row, self-propelled beet harvester.

In a speech heralding the opening of the new Alwalton Showground in 1966, the Society's Vice Chairman, Mr Joe Odam, addressing a large gathering of Show Judges and Council Members of the Society at a dinner held at the Angel Hotel, gave the reasons why he felt certain that the Peterborough Show would go from strength to strength at its new Showground:

The Society has taken a courageous decision in making the move to Alwalton, but it was not their intention to vary the character of the present show. Everyone likes the present site, but believe me we are going to have a better one. Some people say our main ring is the finest in the country and we cannot have another like it, - so we will have a better one. There is also going to be a sportsman's area which will attract specialists to the Show and there will be room for educational visits.

We are going to be successful because we have such an enormous tradition behind us and a tremendous enthusiasm that means we cannot fail.

The geographical situation of the new Showground not only makes it easily accessible from the A1 but it enjoys good parking facilities and competition wise, within a 60-mile radius there is no other comparable show area in existence.

The Alwalton site in fact was not only convenient for the A1 but also very handy for the cross-country east to west A47 and A605 roads. The nucleus of the site consisted of about 250 acres of good arable land. To make this suitable for use as a showground a considerable amount of work had to be done. Hedges had to be removed, ditches filled and drains laid. Additionally the land had had to be ploughed, levelled and seeded with grass and this process took three years, from 1962 to 1964.

The 10th Earl Fitzwilliam played an absolutely crucial role in the negotiations surrounding the acquisition of the Alwalton site and the vacating of the old one at Eastfield. He also arranged for the Peterborough Royal Foxhound Show Society to vacate Eastfield in favour of Alwalton. This meant that the Eastfield site was considerably enhanced in value and that the Society thereby obtained a much better price when the land came to be sold. The Eastfield Showground had consisted of only about 40 acres in total and this gives some idea of just what a massive expansion the Society was now embarking on. It constituted an enormous tribute to the dedication and confidence of its officers, committee members and full-time staff whose hard work and enthusiasm formed the basis of the remarkable strength and drive that were always associated with the Society. All eyes were now fixed on an exciting new future at Alwalton.

Finally, it is worth recording that the area of the Eastfield Showground, now mostly covered by housing, contains the grave of a lion. For many years in the 1940s and 1950s, it was customary for one of the main touring circus companies, such as Billy Smart's, Chipperfield's or Bertram Mills' to visit Peterborough in the winter and to stage its performances at the Showground. One year a lion belonging to Billy Smart's Circus died and there seemed little alternative but to have it buried on the site. This was done in an unmarked grave. Unfortunately the following year it was rumoured that the circus staff were having problems because the animals were proving to be restless and disturbed. Some people put this down to the fact that they had scented the dead creature, amongst them but not of them, as it were.

The Eastfield site was sold to the Peterborough Joint Education Board in 1964 with the proviso that it should be vacated immediately after the 1965 Show.

CHAPTER SEVEN

The Shows at Alwalton 1966 to 1972

So 1966 saw the first Show at the new Alwalton site. The Peterborough Royal Foxhound Show Society had moved at the same time and 1966 was notable, among other reasons, because this was the first occasion on which both the Agricultural Society and the Foxhound Society had the same President, in this case the 10th Earl Fitzwilliam. The Foxhound Society is rather unusual in today's commercialised world in that its activities continue to be unsponsored.

As well as providing invaluable assistance with the sale of Eastfield, the 10th Earl was extremely helpful in the negotiations which surrounded the acquisition of the land at Alwalton. He assisted the Society to obtain Glebe Farm, a holding of 171 acres just south of the A605, then tenanted by Mr R.H. Waterworth. He used his good offices to provide Mr Waterworth with alternative land at nearby Lynch Farm, the tenancy of which had just lapsed while Mr Waterworth also very helpfully agreed to sell the freehold of 22 acres of land abutting onto the A605. The Earl very generously made a further 40 acres available at Alwalton at a peppercorn rent and on the Earl's death, Countess Fitzwilliam equally generously donated this land to the Society. Before this, in 1973, the Society had again enjoyed the munificence of the 10th Earl when he had made available a further 54 acres of land for lease by the Society. In all, 210 acres became available for the Showground and its associated car parks. To mark their gratitude, the Society was later to obtain permission to name their main grandstand, 'The Fitzwilliam Grandstand' while the Foxhound Show enclosure was also to be known in the future as the 'Fitzwilliam Hound Enclosure'. The Society owed a great debt to the 10th Earl and Countess Fitzwilliam and to Mr R. H. Waterworth and his family for making possible the purchase of the Showground.

The Alwalton Showground was planned from the start to take account of the fact that most of the competitors, exhibitors and visitors would arrive by road. Therefore particular consideration was paid to providing effective car-parking and access to the ground for motor vehicles of all sorts. Even with such attention to detail at an early stage, the Society was almost a victim of its own success and traffic congestion in the area of the Show continued to be a bugbear for many years. This in spite of the fact that the car park originally covered 70 acres and had accommodation for 14,000 cars.

The first Show at Alwalton opened on time and with all the necessary facilities available but only by the skin of its teeth! On the Sunday, two days before the

69. Progress at Alwalton

Mr. H.M.T. Jones, Member of Council, and Mr. Roy Bird, Sercretary, cast an appraising eye over construction work at the site of the new Alwalton Showground as the grandstand begins to take shape.

Show, and to a background noise of banging and clattering from toiling workmen, 1,500 people braved the elements for a 20-minute service of dedication. The Bishop of Peterborough, the Right Reverend Cyril Easthaugh, dedicated a plaque which commemorated the opening. Giving a short address, he said that the service celebrated the essentially honourable nature of that fundamental activity, the tilling of the soil and he declared:

> *Nothing is more down to earth than agriculture. Here we see all human activity aimed at one purpose - producing abundant life.*

By way of response, the 10th Earl Fitzwilliam, the Society President, declared that they were seeing the rebirth of the Show and he continued:

> *We wish it all the success it has enjoyed in the past and we know it will go forward to even greater heights on a ground which is among the most modern in this country.*

It was indeed one of the most modern in the country. The Town Planning Department at Newcastle University had been involved with the Society in designing an ideal Showground, starting to plan from absolute scratch. Additionally, Society officers had paid visits to all the other notable Showgrounds of any size in the UK and examined them with meticulous care to see what lessons

70. The Alwalton Showground from the air.

This view taken in a roughly south-eastern direction graphically demonstrates the enormous increase in the size and complexity of the new Alwalton Showground by comparison with its predecessor at Eastfield (see also illustration 54). It also bears witness to the continuing rise in the use of the motor car. In planning this site, it was no longer a major factor to locate the Showground within reasonable distance of a railway station or even of a bus route. A more recent picture taken from the same angle would show housing and other building developments which are now encroaching on land on the left of the picture.

they could provide. It would not be an exaggeration to say that Alwalton was the first Showground in Britain to be designed specifically to meet all the foreseeable requirements of a large Agricultural Show in the late twentieth century and onwards.

The 1966 Show, opening on this extremely impressive new site, maintained its reputation for being at the forefront of agricultural developments and for finding novel ways to inform and amuse the general public. Among the more unusual non-agricultural attractions were displays by Australian axemen and virtuoso performances by the World Tree Climbing Champion. Additionally there was a demonstration of the versatile possibilities of the hovercraft. Although this ingenious machine had been patented by Christopher Cockerell in 1955 and in a simple form shown to the public for the first time in June 1959, it is highly likely that for most of those who watched the demonstration, this would have been their first sighting of a hovercraft. Yet again the Show proved to be the place where new technology made its first appearance in the City.

A pageant was becoming an established event at the Show and this year the centrepiece was a replica of the 'Cat and Custard Pot' public house provided by the now long-closed Phipps' Brewery of Northampton. Loosely based on the novel, *'Jorrocks Jaunts and Jollities'*, published in 1838 and written by R.S. Surtees, the pageant depicted comical fox-hunting characters dressed in the style of the 1820s about which he was writing. Just to ensure that the audience did not take things too seriously, girls in mini-skirts, then the height of fashion and highly attractive on the few girls who possessed legs of the shape and volume to suit them and much less so on the majority who didn't, also took part. Mr Jorrocks, Master of the Handley Cross Foxhounds, was portrayed by Sam Poles, a well-known Whittlesey veterinary surgeon and his huntsman by Mr Robert Van Houtun. All the other characters were pupils of a riding school at Alconbury near Huntingdon, the Fitzwilliam Hunt and its Pony Club and the Huntingdon and District Riding Club. The horses and ponies were unused to ring-work and all this excitement and one threw its girl rider, who was sitting side-saddle. Fortunately she was unhurt.

In all sections except pigs, livestock entries were well up on previous years. Horses, for example, numbered 873 compared to the previous year's 832, cattle 660 as against 601 and sheep 168 as opposed to 132. The weather was unpleasant but that did not deter an estimated 23,000 people who had come through the turnstiles by the end of the second day. No fewer than 5,086 had come on Tuesday, normally not very well-attended by the general public and this was probably partly to do with the novelty of the new super Showground but also because the organisers had rearranged long-established schedules to spread the attractions more evenly over the three days of the Show. Overall attendance was down by something like 3,000 in this year which allowed the Jeremiahs to mutter, 'we told you so', but it was only a short-term trend and especially after the adoption of the 'East of England' name and the series of mergers which led to much larger shows being put on, attendance rose rapidly, exceeding in some years the organisers' greatest expectations.

The well-known Peterborough firm of Peter Brotherhood Ltd whose factory was

71. Pamela - quite a girl!

A feature of the 1966 Show, the first at Alwalton, was the eye-catching 'Pamela'. In fact 'Pamela', the acronym for the 'Perkins Apprentices Mechanically Engineered Lactating Animal', weighed half a ton, had 150 working parts and vital statistics of 8 feet high, 12 feet long and 3 feet wide. Pamela was part of a display called 'Mechanimal Farm' which demonstrated some of the possible applications of automation on British farms. The girl shown in the picture is beauty queen Gloria Bolton of Peterborough. Even though Gloria has her pails at the ready, she would have had a job milking Pamela!

sited close to Lincoln Road in the Walton area of the City, made a welcome return to the companies with trade stands. Their display featured several young craft apprentices operating machines which only the previous month had been featured at Olympia in London in the highly prestigious Chemical and Petroleum Engineering Exhibition. The Mid-Anglia Constabulary also had a stand for the first time on which they displayed a police patrol car and motor cycle, aqualungs and a variety of security devices. Popular with all were the Alsatian dogs, 'Rex' and 'Ingo', taking a break from their more arduous everyday duties.

Peterborough Amateur Radio Society had a short-wave transmitting station at which they proudly displayed a variety of radio equipment, old and new, which included crystal sets made by two of the older members when they had been schoolboys in 1922. An extremely unusual item on display was the only known wireless set in the world made on a sheet of plate glass. This was the creation of a Holbeach man and had been produced about 40 years earlier.

The fact that there were commercial stands with displays by companies as far afield as Scotland, Wales and the West Country was clear evidence of the high commercial standing in which the Peterborough Society's Shows were held. One bonus that came from having the extra space at Alwalton was that for the first time for years, it was possible to allocate accommodation to all those organisations who had applied for stands.

The farm labour force, with its full-time workers diminishing annually by about 6%, combined with the need to continue increasing the output of livestock products, made it almost inevitable that there would be considerable interest in applying ever more sophisticated forms of mechanisation on the British farm. A feature called 'Mechanimal Farm' had a number of interesting displays including in the pig section: slurry tankers, slurry elevators and equipment for organic irrigation and equipment for automatic floor feeding and thermostatically controlled ventilation. Another item of interest were the tall 'cropstores' or silos which have since become commonplace in many parts of the British countryside and which are generally used for storing grain and forage and designed for maximum conservation efficiency. Real live animals on 'Mechanimal' Farm included 'Jersian' heifers, the first cross-breed by a Jersey bull out of Friesian cows, and Charolais crosses.

The Wednesday of the Show was a big day for Frederick Sidney Hurst. He received a long service medal from the 10th Earl and Countess Fitzwilliam in grateful recognition of his 40 years as a cowman working for Horrell's Farmers at their Westwood, Peterborough, dairy. Mr Hurst started his employment with Horrell's on leaving school and although primarily employed as a cowman, he recalled that he had been happy to turn his hand to other jobs that needed doing and had worked on occasions in the bottling plant and had even gone out doing doorstep deliveries. His tenacity and loyalty evidently ran in the family because he told the *Peterborough Advertiser* that he had four brothers, two also working at Horrell's and two employed on the railway and not only had these four accrued about 150 years service between them but like him, each only had the one employer.

A correspondent for the *Peterborough Standard* provided an excellent flavour of the 1966 Show as seen through her own eyes as a wife and mother. She explained how her enjoyment began as soon as she passed the entrance to the Showground and was assailed by a mass of colourful stands all of which beckoned her to come closer and consider what they had on offer. No sooner was she in the Showground than, with a clap of thunder, the heavens opened and she had to dive for cover.

Among the first stands she visited on her tour was that of Carters Tested Seeds Limited who displayed a number of novel vegetables, some little known at the time, including kohl-rabi which she was assured was a more than adequate substitute for cabbage and which should be prepared and served like turnips, a yellow tomato that had a much thinner skin than normal red tomatoes and, perhaps bemused by this cornucopia of vegetable delights, she was offered and carried away with her, a purple runner bean which, it was claimed, would, when cooked, appear even greener than the normal green variety and taste every bit as good.

Anything that would minimise household drudgery was of course of immediate interest and she made a beeline for the stand giving a demonstration of the 'Kenwood Chef'. This efficient and extremely economical, mixer-cum-mincer-cum-liquidiser would, she was firmly assured, instantly and effortlessly dismiss all the donkey-work and tedium from food preparation and leave any housewife immaculate, radiant and full of bonhomie to greet her children and the man in her life when they returned from their toils to the family home in the late afternoon. She marvelled at the sight of the finely sliced cucumber and delicate shreds of carrot that were produced so expeditiously. She then sipped, with understandable caution at first but then with growing relish, an invigorating, nutritious and very economical drink, composed of lemon, apple, sugar and water which was created in an instant by this veritable polymath of a kitchen gadget. Nearby was the 'Drimaster', a sensational dishwasher which, it was claimed, washed, rinsed and dried in four minutes and thereby opened up tantalising visions of an instant escape from the everyday grinding irritations of filthy tepid dish water and congealed grease that have been the bane of life for all those who have worked in kitchens throughout recorded time, and probably before.

Elsewhere, she was wooed by the siren calls of new vacuum cleaners, including one with a cunning device that turned it from a cleaner of carpets into an ultra-efficient floor-brush at the flick of a switch and which called forth this comment:

> *How many of these marvels I saw and wished for - ours only for the money - and the domestic thread glowing golden at the thought of a kitchen expertly run by electrical gadgets, in a home beautifully cleaned by electrical gadgets, and one's housework schedule speeding efficiently from one switch to another.*

Our correspondent moved on, carrying her enthusiasm with her, to the display put on by the Singer Sewing Machine Company. Here she spoke lyrically of the scaled-down sewing machines that would allow girls, aged little more than ten, to absorb the basics of their use and then be able to graduate in due course to the much more sophisticated operations which could be carried out with the 'Swing Needle Model'.

Eventually all this dilly-dallying took its toll and she found that she simply did not have the time or the stamina to get round and see all the exhibits that interested her. As she said:

> *I wanted to visit Nestle's to see their demonstration kitchen. I wanted to question an authority on potatoes about this staple item going black before cooking - as it disappointingly does on occasions. And I wanted to see ... the British Egg Marketing Board (where 'add an egg and cook still better', was the theme of the cookery demonstrations), as well as the particularly interesting stand of the British Sugar Corporation.*

She completed her day with a visit to the Women's Institute stand and the Flower Show. She was clearly delighted with the W.I. contribution:

> *Each W.I. Branch competing had to arrange the background to its entries and this produced some very attractive settings. Elegance and freshness seemed to be the tone of most, but an appealing one...was a background clearly showing the Market Place and the entrance to the Cathedral. I loved this and think that it would give a lot of pleasure when upon seeing it folks would immediately recognise the familiar scene.*

Moving on, battling against time, to the Flower Show our doughty correspondent met up once more with the W.I. She was greatly impressed by the artistry and technical ability of the W.I. entrants' floral displays and described the art of the Flower Show as a whole as 'breathtakingly lovely'. Having declared that while the weather had certainly been dreadful, it had not taken away any of the pleasure of the visit, she concluded:

> *... as I left the ground I still heard in retrospect the music of the military and pipe bands of the Brigade of Gurkhas, saw the sculptured beauty of hunters being exercised by their riders, and recalled snatches of conversation heard throughout the day on a patchwork of topics, including horses, farming, Swiss governesses, fashion, children, processes for this and that, and cars.*

It was evident to the Society's officers and leading members that regional Agricultural Societies, while continuing to stage and put much of their effort into their Annual Shows, needed to take a much wider role in agriculture if they were to justify their continued existence. Never ones for simple pious resolutions, the Peterborough Society took a major step towards implementing a more proactive approach when, in 1966, in collaboration with the Yorkshire Agricultural Society, they developed a farm secretarial service called Farm Records (Eastern) Limited. This provided a service for farmers from the North East of England down through the Midlands and East Anglia as far south as the Home Counties.

Later in 1966 an even more important development was the formation of the Agricultural Development Association (ADA) in conjunction with three other forward-looking and progressive Agricultural Societies, namely, the Bath and West and Southern Counties, the Shropshire and West Midlands and the Yorkshire. The ADA's main purpose was the encouragement of activities that would assist the spread and implementation of the results of scientific research into everyday farming practice. The key to the success of ADA was to be collaboration between the research work being done in the universities, the teaching and dissemination of good practice in the agricultural colleges, the research and commercial drive of industry, and the farming community at large.

In 1967 the talk of the agricultural world of mid-Anglia and the south-east Midlands was the discussions that took place in the spring between the Peterborough Society and six neighbouring agricultural societies. These were the Huntingdonshire, Bedfordshire, Cambridgeshire and the Isle of Ely, Leicestershire, Northamptonshire and Rutland Agricultural Societies. The talks started off purely informally but as common ground was found and progress made, the intention was that the outcome of the talks should be kept secret until an agreed joint statement could be produced.

The word 'secret' only has to be mentioned for wild rumours, often of the most far-fetched and speculative nature, to start circulating and then developing a life of their own. Usually those who know least, say most. So, in August 1967, to scotch continuing speculation, Roy Bird, Secretary of the Peterborough Agricultural Society, issued a statement to the effect that the meetings had involved a full and frank exchange of views about the problems facing societies and the difficulties of organising agricultural shows and how these could be overcome. Some members believed that these discussions were actually the first step towards a possible merger of two or more of the region's leading agricultural societies and it was clear even at this stage that any such moves could open up a Pandora's box of understandable fears among the societies of losing tradition, local identity and organisational autonomy.

Many predicted that the 1967 Show would be a record-breaker even before the turnstiles opened. The numbers of both livestock entries and, contradicting the prevailing trend, trade stands, established a new record. A number of modifications had been made to the schedule for the three days and Tuesday when the Royal visitors, The Duke and Duchess of Gloucester, were due, was now scheduled as pedigree day with the judging of all the pedigree cattle, pig and sheep classes. The Foxhound Show was also moved to the Tuesday thereby enabling those interested in hunting to see the judging of both the hounds and hunters on the same day. The journal, *Farming World* for 13 July, 1967 commented:

> *The increase in the number of trade stands - 320 compared with last year's record of 315 - will be viewed enviously by many show societies. In part it would seem to be a case of success breeding success.*

This view was reiterated by the then Society Vice-Chairman Mr Joe Odam who was full of praise for the way in which everybody concerned had worked so hard to recreate the best of the atmosphere of the old Eastfield Showground which was now combined with the superior facilities that were on offer at Alwalton. He said:

> *We can at least claim that at a time when agricultural shows are feeling the pinch, this show appears to be more than holding its own.*

The first day of the 1967 show had seen heat of almost tropical intensity but this gave way to milder conditions on the Wednesday and, with the typical capriciousness of the English weather, a thunder storm late in the day. Thursday was merely overcast. Of interest was the fact that some visitors to the Show, including a number from Europe, flew over the ground in a variety of aircraft from single-engined Austers to twin-engined Skymasters, before availing themselves of the new airstrip at nearby Sibson. How many regional Agricultural societies at this time could boast international visitors arriving by plane? Two unfortunate incidents marred the show. In one case a woman attendant at a trade stand was violently struck on the head by a metal fitting which had detached itself from a cut away parachute during a free-fall display by men of the Parachute Regiment. She had to be hospitalised with head injuries but made a full recovery. The other was a raid by sneak-thieves who stole about £50 worth of goods from the stand exhibiting goods of the well-known Peterborough firm of Robertsons All Sports.

The outcome of the discussions taking place that year between the various societies was that the Bedfordshire, Cambridgeshire and Isle of Ely, Huntingdonshire and Peterborough Societies would come together to put on a regional show in 1968, this show taking the place of the separate shows normally staged by the four societies. This joint venture was given the indicative title, 'The East of England Show'. A managing committee with nominees from each society and chaired by Joe Odam, as Vice-Chairman of the Peterborough Society, would plan and run the show but the participating societies would maintain their separate identities. The societies had a total membership of 7,000 to 8,000 and felt that there was strength to be had from collaboration on the staging of the show even if they continued separately with their non-show activities. This first regional show was underwritten by the Peterborough Society.

The Show Management Committee held on Monday 22 April consisted of the following members: Mr J. Odam (Chairman); Messrs J.F. Bance (Huntingdonshire); J.A. Clayton (Cambs and the Isle of Ely); R. Cullen (Hunts); A. Deptford (Cambs and Isle of Ely); K.G. Fryer (Peterborough); W.T. Godber (Beds); A.B. Gould, S. Harris, M.W. Jeffrey, H.M.T. Jones (all Peterborough); P.L. Lenton (Huntingdonshire); G.P.L. Little (Peterborough); A.T. McCreadie (Hunts); F.C. Mitchell (Peterborough); R.J. Parrish (Beds); R.M. Peacock (Beds); L.H.E. Perks (Beds); D.W. Reeves (Beds); W.R. Riddington, C.E.W. Saunders, J. Scorer Jnr; J.N. Scorer (all Peterborough); F.G Starling (Cambs & the Isle of Ely); H. Waterworth (Peterborough).

Also in attendance were the Secretaries of the respective Societies: Messrs G. Beazley (Beds); R.W. Bird (Peterborough); R. Keene (Hunts); A.E. Vyse (Cambs and the Isle of Ely) and Assistant Show Secretary J.A. Wright.

Apologies had been received from Messrs S.H. Gould; J.R. Horrell; D.O. Powell and C.D. Tebbs. This collaboration was taking place as part of a wider process of discussion among senior officers of these four Societies around the question of some kind of amalgamation and the effects that would have on the Societies individually and on agriculture in the region. The Marquess of Exeter who was

President at the time, played a very important role in the talks around the difficult question of mergers. He was able to command from all the parties concerned and brought his considerable persuasive powers and what are now called 'inter-personal skills' to bear on managing what might otherwise have been quite tortured discussions. Kenneth Fryer, as Chairman, made many absolutely invaluable contributions as leader of the Peterborough Society's delegation, not least in recommending that the first Chairman of an amalgamated Society should be the highly popular and effective Peter Lenton, Chairman of the Huntingdonshire Society, at that time the smallest of the Societies engaged in the discussions. Credit must also be given to the senior members of each of the participating societies for the very positive way in which they approached these sensitive negotiations. Despite suggestions in certain sections of the media that this was a 'takeover' by the Peterborough Society, those involved in the negotiations knew that it was in fact, in the real sense, a merger.

The situation facing agricultural societies was put very succinctly by Roy Bird, the Secretary in a paper in 1967 entitled, 'The Future of Agricultural Shows'.

Agricultural Shows have been the subject of wide and varied comment over the last 20 years by journalists, farmers, manufacturers, agents, sportsmen and countryfolk in general. Many articles have been written expressing differing and controversial opinions. Permanent Showgrounds have played an important part in Society policies. All the unwieldy travelling exhibitions are now anchored. To some extent a pattern has emerged. The Royal Show is now firmly established on its permanent ground at Stoneleigh Abbey in Warwickshire and will shortly be joined by the British Horse Society which is to establish a National Equestrian Centre there.

...The picture for regional and County Shows is not so clear. Trade support was highlighted in recent years by the withdrawal of manufacturers and dealers from some Shows and this alone may well dictate the future of many summer Shows. It is difficult to visualise Shows without the traditional displays of machinery, merchants and allied industry stands. For years, manufacturers and agents have been saying that Shows should get together... Already one large amalgamation of Societies has taken place in the South of England with the Royal Counties, Sussex County and Tunbridge Wells. I feel members of the trade believe there is a future for Agricultural Shows which are prepared to keep abreast of modern developments. Societies must accept change, work closely with Trade and Livestock Exhibitors and provide Showgrounds with all modern conveniences and amenities. Few people connected with farming wish to see the great tradition of British Agricultural shows disappear. The large summer exhibitions are the only opportunities for farmers, their managers and workmen to see the latest developments in farming techniques. It is accepted practice that before purchasing new equipment or changing a system, one should first visit a Trades Exhibition where on the spot comparisons can be made followed by a field demonstration.

The support of the farming community and the Trade is essential, but few Shows can now exist without a large gate of townspeople. There are insufficient farmers to provide necessary financial support for Shows, particularly in the face of the high cost of staging these events, and it is more than ever essential that a larger attendance of town people is attracted to provide adequate gate income. It is however of paramount importance for a Show to maintain a correct balance to keep the interest of the Agriculturalist.

Farmers are by nature, conservative in outlook, but if the Show becomes virtually a Fair in order to attract large gates of townspeople then the Agricultural flavour can easily be lost together with the support of the farmers. No Society should miss the opportunity of selling British Agriculture and British Farm Foods to Town housewives who visit the Summer Shows. The Farm Food Fair at Peterborough is a main feature of attraction.

Apart from the importance of Shows to Agriculture, there is another important and interesting side - the social one. The first day at any Regional or County show is the greatest social occasion for that particular area. Farm workers now take more and more interest, as indeed they should, in the progress of their highly specialised business. I am sure that many farmers would never think of purchasing equipment without first discussing it with their key men.

For Regional Societies it has, in recent years become increasingly apparent, that in addition to staging the Annual Show, Societies must widen their activities. These societies should, and can, take a far more active part in farming generally. Far too many of them are inclined to concentrate purely on the Summer Show... their grounds must be put to good use throughout the whole of the year. Depending on the location of the ground there may well be opportunity for trade demonstrations - fertilisers and machinery, pedigree stock shows and sales and many non-agricultural activities - all these must be investigated... There is a future for shows, but societies must be run on professional lines and meet the high standards demanded by farmers and the general public.

Between the 1967 and 1968 Shows, the Edmund Beck Memorial Bandstand was moved from Eastfield to a new location at the Alwalton site and a new judging box for show rings 1 and 2 had been built and generously paid for by the Horrell family as a memorial to Mr Harry R. Horrell who had been a member of the Society's council from 1935 to 1963. Another landmark was the Robert Bibby Memorial Clock. This had been placed on the tower of the ornamental gateway to the old Eastfield Showground. It was now located near the entrance to the Alwalton Showground where it stood like a sentinel until it was dismantled and then re-erected on the East of England Agricultural Society's Administration Centre. In 1971 a plaque commemorating the clock was unveiled by Prince Philip, The Duke of Edinburgh and in 1998 a brand new plaque was to be placed on the Administration Block where it could be easily seen by visitors to the Show and the other activities held on the site. Also indicative of the Society's developing importance was the move in 1968 from its familiar and long-established offices at Priestgate in Peterborough City Centre to the new Administration Centre at Alwalton. The Priestgate office had simply been outgrown and therefore could no longer provide an effective headquarters for the Society.

The 1968 Show in its new and enlarged form generated a considerable amount of interest and eyes were on it, not only locally and regionally but throughout the agricultural community of the whole of England. Was the new format going to work? Were the four societies working closely together stronger than they would have been individually? What would be the response of the competitors, the trade exhibitors and the paying public?

Direct comparisons with the various constituents' previous Shows would be meaningless but certainly the first East of England Show was a great success in attracting a significant increase in trade stands and livestock. Unfortunately the weather was appalling which undoubtedly deterred paying visitors. After a prolonged period of dry weather, four inches of rain fell on the Wednesday and Thursday preceding the show, accompanied by unusually high winds, and together they flattened and destroyed all but one of the marquees. The marquee which remained standing was that belonging to the NFU and it did so only because some of its members had kept an active vigil all night and carried out repairs when required.

Another inch of rain fell on the Sunday but Herculean efforts on the part of the staff, additional workers brought in from nearby farms and the contractors meant that the show went ahead as scheduled despite these conditions. A total of 61,940 people visited the Show which in the circumstances was remarkable and was considered by all concerned to be a great success. One aspect of collaboration which was clearly vindicated was that such unfavourable weather conditions would probably have proved disastrous to any of the participating Societies had they been staging their own individual Shows. Two examples of fraternal goodwill, arising from the atrocious weather, were the loan of a variety of show equipment free of charge by the Yorkshire Agricultural Society and of 20 Landrovers from the Royal Society.

The Minister of Agriculture at the time, Cledwyn Hughes, fulsomely described the Show as 'a glittering shop window' for agriculture and he praised the organisers' decision to amalgamate their individual county shows. He continued:

> *From what I have seen so far it is quite clear that this new venture is going to be a very great success and of considerable importance to agriculture in this area, which is without question one of the most important agricultural areas in Great Britain.*

He was sure that the East of England Show was going to be one of the country's major agricultural events.

To give a flavour of what the 1968 Regional show had to offer, the three days not only had 300 exhibition stands and 3,000 livestock entries but also included demonstrations of agricultural machinery, a display and market of British farm produce, show jumping, the Foxhound Show, floral displays, a flower show and a feature containing vintage farm machinery. The number of livestock entries was very encouraging given the foot-and-mouth epidemic and problems associated with the eradication of brucellosis. Especially for the ladies was a brand new feature, 'Demotique'. As well as numerous cookery demonstrations, fashion and hair styling presentations every day, guest appearances were made and among the stars who appeared was Margaret Patten, generally recognised at the time as the country's top cookery expert. In addition there was an art exhibition, a dog show, clay pigeon shooting, a W.I. stand, go-karting and a tug-of-war contest while major spectacles included ballooning, military massed bands and displays, a Horse and Hound Pageant, motor cycle racing, driving classes, a hovercraft and a man being shot from a rocket cannon. Here was the Agricultural Show with something for everybody.

The first East of England Show was certainly a great success despite the efforts of the weather to sabotage it. It had demonstrated beyond dispute that this enlarged show based on a permanent ground at Peterborough had the drawing power and the amenities to attract trade exhibitors, livestock entries on a large scale and a host of other exhibits, demonstrations events and activities to entice the fee-paying public through the turnstiles.

Times were definitely hard for the smaller Agricultural Societies. An ever-increasing range of counter-attractions in a world of rising incomes and growing time for leisure were constantly being advertised and were competing with whatever Agricultural Societies could offer. It was a tough but inescapable fact that the shows staged by the smaller societies were increasingly losing money, caught in a spiral of rising costs and shrinking revenue and profits. Businesses had to make increasingly tough decisions about how, when and where to promote their products in a distinctly harsh economic climate. The continuing success of the Royal Agricultural Society's Shows was clear evidence that for survival, 'Big was Beautiful'. There were sound economic reasons why businesses should prefer to exhibit at one very large Show rather than half-a-dozen small ones.

Clearly all the organisations and individuals that came together to stage this Show were keen to identify the advantages that accrued by staging one combined show rather than the four separate ones. Mr Gilbert Beazley, Secretary of the Bedfordshire Agricultural Society, aptly summed up the general feeling. He hailed the event as a great success and said:

> *The amalgamation was undoubtedly worthwhile. The show was a great success despite the weather. We attracted a lot of practical farmers and were able to give them what they wanted. With this summer show run hand in hand with our other activities, I think we shall give the agricultural community what it has been looking for.*

On I February, 1969, the Huntingdonshire Agricultural Society, formed in 1837, merged with the Peterborough Agricultural Society to form the Peterborough and Huntingdonshire Agricultural Society, a move that was felt would be of mutual benefit and would also further the interests of agriculture in the area. The Long Sutton and District Agricultural Society, also founded in 1837, decided to become affiliated but it did not actually join or contribute funds to the new Society.

1969 maintained the tradition that shows at Alwalton were gaining for combining all that was best about the old-fashioned agricultural show with evidence of the latest innovation and enterprise. A major farming feature was the 'Pigs in the 70s' demonstration staged by the Meat and Livestock Commission, the Ministry's Food Division, the Agricultural Development Association and the Show's own staff. Exhibitions of interest to farmers and general public alike were on the themes of 'Soil Structures' and 'The Change to Metric' as well as the 'Potato Fayre' put on by the Potato Marketing Board.

There was some reduction in livestock entries with cattle, pigs, sheep and heavy horses down slightly but a rise in the light horse, pony and goat classes and an increased entry total for the dog classes exceeded the organisers' bravest hopes. Horse lovers had much to enthuse over with a total entry up by nearly 100 on the previous year. Hound shows were held on all three days. On the Tuesday they featured the Master of Bassett Hounds Association, on Wednesday the Peterborough Foxhound Society and the Master of Harriers and Beagles Association on Thursday.

In the Main Ring was staged a 'Grand Rural Montage' with over 500 participants and which included an Edwardian Royal Ascot scene, a tractor activity drive, about 20 Young Farmers Club decorated vehicles and vintage cars. This all culminated in a 'beating retreat'. Thursday featured a clay pigeon shoot organised by the Whittlesey Wildfowlers' Association while events during the three days in the main ring included a military and massed band display by the Royal Anglian Regiment, a tank 'musical drive' by the Royal Armoured Corps and various other displays by the Royal Air Force Queen's Colour Squadron and the United States 17th Air Force Band. There were the well-established and ever-popular attractions such as top-class show-jumping, the W.I. market, the NFU stand, craft stalls and the reappearance of the highly successful 'Demotique' demonstrating a number of skills concerned with hair-styling, cookery and fashions. For the children there were a full-size replica of an Apollo space capsule and two bogus but celebrated motor cars. These were the original 'Chitty Chitty Bang Bang' that starred in the musical of that name and Lady Penelope's pink Rolls-Royce, 'FAB 1', from the Thunderbirds TV series. It was entirely appropriate that this bizarre vehicle should be exhibited at Alwalton, given that Sir Henry Royce was born in the locality.

The previous year considerable publicity had been given to the fact that certain visitors, including a number from overseas, had flown to the Show in their own machines, landing at the Sibson airfield owned by Charles W. Shelton Ltd. In 1969, building on the previous year's events, an air rally of these visitors and their light aircraft was organised and constituted another first for the East of England Show.

The Show activities which have been mentioned here were only a small sample of what the three day Show offered but provide clear evidence that the 'East of England' had a major role not only as an agricultural exhibition serving the industry by promoting good practice but that it was also a significant public relations exercise helping the visitors to gain some insight into the nature of British agriculture and its activities. At the same time, a Show was being put on that appealed to a family audience and could therefore generate vital income from paying visitors at the turnstiles.

Irrespective of the best efforts of the organisers, the weather will always have a major bearing on the success or otherwise of each and every Annual Show. With the rain having done its very best to drown the 1968 Show, in 1969 it was as if the sun, determined not to be left out of things, went out of its way to bake everybody and everything! While the organisers must have been thanking their stars for the sun and blue unclouded skies which attracted over 6,000 more visitors than the previous years, the temperature at one stage soared to 95 degrees in the shade! This caused great distress, most of all to the pigs, of which there were more than 200 entries. The order of events had to be changed so that the largest pigs which

72. The Royal Visitor, 1970.

HRH The Duke of Edinburgh at the 1970 show with Joe Odam, that year's Show Chairman and sponsor of this definitive history of the Peterborough Agricultural Society.

73. A Royal Visitor at the Livestock Judging.

A feature of the Shows of both the Peterborough Agricultural Society and its successor the East of England Agricultural Society, has been the regular attendance of one or more members of the Royal Family. Here we see HRH The Duke of Edinburgh viewing the prize pigs in the judging ring with Mr. Sam Harris (Steward), a future Society Chairman.

74. Top Goats, 1971. A line up of winners at Alwalton.

75. 'History in Action.'

The controlled use of steam power in traction engines and locomotives exerts a perennial fascination on young and old alike. Here the picture shows a demonstration of a steam powered threshing machine at the East of England Show, 1972.

were most at risk from the heat could be judged before the sun reached its zenith. Through the Tuesday in particular, farmers and herdsmen worked frantically to keep the suffering animals cool by spraying the canopies above the pigs with water and by rubbing the pigs with vinegar. As Mr Hal Thompson, the Regional Livestock Officer of the Meat and Livestock Commission told the *Peterborough Standard*'s reporter: *Vinegar is to a pig like smelling salts are to a human being but no-one knows why.* Despite their best efforts, sadly a number of pedigree animals succumbed to the heat and died of heart attacks. With the number of trade stands showing a significant increase over 1968 and paying visitors up, it was clear that the benefits to the partners of working closely together on the Show had not been ill-conceived and were not just a flash in the pan. The Show continued to move forward.

On 1 April, 1970 the Peterborough and Huntingdonshire Agricultural Society went on to merge with the Cambridgeshire and Isle of Ely Agricultural Society to form the East of England Agricultural Society and one year later to the day, the Bedfordshire Agricultural Society (founded 1801), also joined. In 1970 the Society was credited with yet another national first when it staged a Conservation Exhibition at the Show. This proved an immediate success and has continued to this day, remaining a popular feature.

76. The Wild West comes to Town.

A picture of an action-packed Main Ring with the Stampede at the 1972 Show.

77. Royal Naval Helicopter Display.

The crowds appreciate the versatility of the Royal Navy as this helicopter goes through its paces at the 1972 Show.

78. Swashbuckling stunts at Alwalton.

Daring manoeuvres by the Royal Marines Motor Cycle Display Team at the 1976 Show. Although some purists argue that displays of this sort, excellent though they are, have little or nothing to do with Agriculture, the fact is that they provide the kind of entertainment that attracts people through the gates. This generates income which can then be used by the Society in its many activities in support of Agriculture.

In 1971, the attendance at the Show reached what was then an all-time record figure of 85,776, which was an increase of 18,000 over the previous year. This success could be seen as an unequivocal endorsement of the move towards a regional Agricultural Society.

1972 saw a further and final merger, in this case with the Northamptonshire Agricultural Society which had been founded in 1848 and with this addition the East of England Agricultural Society became the Society which it is to this day. By a happy coincidence, attendance figures topped 100,000 for the very first time in this momentous year in the history of the Peterborough Agricultural Society, which, in 1797, had started with a brief to serve the immediate locality but which had now, through its own efforts and those of the Societies with which it made a series of judicious mergers, become a vital constituent in one the country's foremost Agricultural Societies.

CHAPTER EIGHT

The East of England Agricultural Society at Alwalton

Although the Peterborough Agricultural Society now no longer exists in name, for the sake of historical continuity, the authors felt that they should provide a brief outline recording some of the more important developments that have taken place at Alwalton under the auspices of the East of England Agricultural Society.

79. The Royal Visitor, 1976.

H.M. Queen Elizabeth, The Queen Mother strolls around the Alwalton Showground with the Chairman of the Society, Mr. H.M.T. Jones (right) and Roy Bird, the last Secretary of the Peterborough Agricultural Society.

80. The 10th Earl and Countess Fitzwilliam at the Show, 1977.

The Fitzwilliams show their continued support for the Society as their ancestors had done before them.

Speedway racing was introduced in the 1970s and has helped to provide extra income for the new Society by making use of the ground and its facilities at times other than when the Annual Show and major agricultural events were being staged. 1973 saw the first 'Expo Steam and Country Fair' organised and staged in conjunction with the *Peterborough Evening Telegraph*, this event taking place on the August Bank Holiday weekend. As well as useful revenue, these events offered facilities to the citizens of Peterborough, provided publicity for the Society and helped to add to the fame of Peterborough nationally and even internationally at a time when the City was starting its expansion as a designated New Town.

Although 'Expo Steam' eventually folded, speedway continues while other annual events have been developed including the British Motorcyclists Federation Show and 'Truckfest'. These have not only been appreciated by the people of Peterborough but likewise have drawn visitors from across the country and from overseas. The Alwalton Showground is in use for something like 200 days of the year and the Calendar of Events for 1998 included such diverse activities as the East Midlands Canine Fair, the Tug-of-War National Championships, the

81. HRH The Duke of Gloucester at the 1977 Show.

The Duke takes time to examine the workings of a sectionalised diesel engine on the Perkins Engines Company stand.

National Shire Horse Show and the Ponies Association UK Show. Clearly the City's economy benefits by the spending of those who participate in these and the innumerable other events that take place at Alwalton.

In 1978 the show was visited by H.R.H. The Prince of Wales who wanted a competition to be established which would encourage young people to take up rural crafts and skills. The Society therefore has made a trophy available to be held by the recipient for a year and additionally a cash prize which was initially £500. The first winner of the of what was to become a prestigious award was Stephen Halford, apprentice saddler to his father at Crowland, near Peterborough. He gained the prize for a first-class show bridle and saddle for a heavy horse. In recent years it has proved quite difficult to find sufficient entrants of the appropriate calibre and the scheme is currently being re-examined with a view to finding ways of encouraging more interest.

In 1978 'The Coates Barn' project was completed. This involved removing an ancient and largely derelict thatched tithe barn from a site deep in the fenland at Coates near Whittlesey. It was transported to Alwalton and then re-erected on the Showground, using original materials as far as possible, thereby providing a

82. HRH Princess Alice at the 1977 Show.

HRH Princess Alice strolls round the Show and mingles with the crowd. Note the priceless expression on the faces of the twin girls on the right.

permanent reminder of past agricultural and building practices in the area for long served by the Peterborough Agricultural Society.

In 1980 the Society started a newspaper, the *East of England Journal*, published six times a year and circulated in-house, being one of the first Agricultural Societies to do so. This has now been changed into the *East of England Magazine*, the evidence being that readers generally take the content of magazines more seriously than they do that of newspapers. Again the Society has proved to be at the forefront of developments because only a handful of similar bodies produce a magazine at the present time.

In 1983 the East of England Society collaborated closely with the Royal Agricultural Society to stage a National Cereals Demonstration. This was held at Cambridge and was called 'Wheat 83'. Over 8,000 farmers and others with an interest attended this initial event which has become an annual and highly successful event.

83. HRH Prince Charles at the 1978 Show.

The Peterborough and the East of England Agricultural Societies have always enjoyed an excellent relationship with the Royal Family as a whole. Many of its individual members, including those in the present day, have had a keen and wide-ranging interest in agricultural and rural matters. The presence of Royalty at the Show helps to attract visitors and boost revenue.

Each year the Society runs a Farms Competition, sponsored by Sidney C. Banks plc, Grain Merchants and Seed Specialists based at Sandy in Bedfordshire. Originally, it was open only to farms in Bedfordshire, Cambridgeshire and Northamptonshire but it has now been broadened to include any others located within 50 miles of the Alwalton Showground. This competition provides recognition and publicity for examples of current progressive and economic husbandry which are environmentally sensitive.

In the early 1990s, the decision was taken to collaborate with the Royal Agricultural Society of England to establish the Sacrewell Arable Farm Research Station at the William Scott Abbott Trust Farm at Thornhaugh, just a few miles up the A1 from Alwalton. This venture has now become a well-used popular resource and one which most schoolchildren from Peterborough and the surrounding areas visit and where they can observe a wide range of agricultural techniques and farming memorabilia.

The excellent collaboration that the Peterborough Agricultural Society enjoyed with a wide range of outside organisations has been continued and expanded under the umbrella of the East of England Agricultural Society. Relationships with the Royal Agricultural Society of England have been mentioned but locally, and nationally, the East of England Society also works well with, for example, the National Farmers' Union, the National Union of Agricultural and Allied Workers (now part of the Transport and General Workers Union), the Young Farmers Clubs, Women's Institutes and many other organisations with an involvement in rural life. Further evidence of practical and valuable collaboration was the creation of the East of England Farm Scholarship in 1996. Here the Society enjoys the partnership of Reading University and Barclay's Bank. The scheme was launched based on the perception that insufficient graduates of high intellectual calibre were being attracted to make a career in the broad sphere of agriculture. In particular, graduates with good first degrees were often unable to afford to continue their university studies at Masters' Degree level. This Scholarship provides a sum of £20,000 to allow a student to partake of a year's postgraduate study in order to obtain a second degree and it also supports one year of work experience. Application for this sponsored Scholarship is open to all with the requisite academic qualifications.

The National Beet-Harvesting Event is another example of mutual co-operation where the Society and British Sugar plc jointly staged the National Beet-Harvesting Event in October 1998. The Society is strongly committed to initiatives of this sort and is constantly looking for new ways in which it can provide 'hands-on' practical support for the agricultural community. In so doing it can proudly say that it remains true to the ideals of its founders back in the late Eighteenth Century in continuing to encourage innovation, the spread of ideas and information and practical support for the farming fraternity.

The years during which the East of England Agricultural Society has had its splendid site at Alwalton have not been easy ones for the farmer and the agricultural community as a whole. This period features a large number of consolidations of land-holdings, reductions in the numbers both of farmers and agricultural workers and large numbers of mergers not least among companies in such associated activities as agricultural machinery manufacturing, dealerships and the supply of such requisites as seeds and fertilisers. Agricultural machinery is becoming more sophisticated and expensive, now making use of computers and robotics and consequently demanding greater skills from the remaining workforce. It is to be hoped that the job security of agricultural workers becomes greater in the 21st Century and that their financial rewards and status rise in accordance with the ever greater occupational skills they are required to demonstrate.

Older Society members understand and many regret how a variety of factors have led to a much less personal way of doing business. A great and congenial tradition among farmers and their suppliers in the Peterborough area used to be a visit on Saturday morning to the Cattle Market, adjacent to the City Centre, followed by attendance at the Corn Exchange and then the clinching of deals and socialising at the 'City and Counties Club' or perhaps the 'Still' pub in Cumbergate. In this way, a network of mutual trust and support based on face-to-face meeting was built up and it is clear, if regrettable, that this way of conducting business has gone for good, replaced largely by telephone calls to suppliers perhaps 200 or more miles away.

In agriculture it is very much a case of plus ca change. Britain's economy and her social fabric having undergone a massive transformation during the 200 years of the existence of the Peterborough Agricultural Society and its successor, the East of England Agricultural Society.

Currently, two major public relations and education tasks confront the East of England Agricultural Society and, indeed, Agricultural Societies everywhere. The first is to keep the general public informed about the vast range of food products that are grown, raised or reared by Britain's farmers and to emphasise the consistently high quality of these products. The second is to explain that those involved with farming know the countryside as well as any and can therefore be said to be among the best custodians of the countryside and the rural way of life.

The Peterborough and East of England Agricultural Societies have been at the forefront of change and progress in the period under review. We can be sure of little else in the next 200 years except that change will continue to be a constant factor. We cannot possibly foresee the variety or scale of the changes that will affect agriculture in that period. However through all this maelstrom of transformation, the basic task of those engaged in farming has been and will continue to be the same. The East of England Agricultural Society can be pleased and proud of its heritage and can look forward confidently to the next 200 years.

CHAPTER NINE

'Three of the Best'

The Peterborough Agricultural Society was extremely fortunate since the latter part of the nineteenth century in being served by three exceptionally able Secretaries who stayed in office for long periods and guided the Society through crucial periods in its growth. These men did a job that was not particularly glamorous, that called for organisational skills of the highest order and where they were first in line if anything went wrong especially at the Society's annual highlight, the Summer Show. It could be argued that a vital part of the success story of the Peterborough Agricultural Society is provided by the contributions made by James Little, Robert Bibby and Roy Bird.

James Little

Not much has been recorded of James Little who came into office as Secretary of the Society in 1874 when Britain was still the 'Workshop of the World' and Agriculture was entering its 'Great Depression'. He was a countryman born in Lincolnshire and, although physically small, was wiry, strong and extremely energetic. He came of farming stock but had no wish to enter farming directly himself. However he had a detailed knowledge of agricultural affairs gained not only through his upbringing but also through his business as an auctioneer and valuer in Peterborough.

His was the longest serving Secretaryship at 45 years and it is highly likely that he would have completed 50 years had the Society's Council not decided to make a double blow for youth by ousting him and the Chairman, Alderman W.G. Moore who had 21 years service in that capacity, at the Annual General Meeting in 1919. He was greatly distressed to be told that his services were being dispensed with and was hardly mollified on being informed that he was to receive a pension equivalent to half of his salary for three years. It only added insult to injury to be assured that in the event of his death, his widow would receive the pension, but only within the three stipulated years.

James Little was to have the last laugh on some of his detractors because he died in 1934 at the grand old age of 91 and he certainly outlived two of them, Alderman Maxwell and Alderman Griffin.

Robert Bibby

'Bob' to friends and he had lots of them, became Secretary of the Society in 1921 and held the position to great effect until his sudden and untimely death in 1957 at the age of 70. He was a Lancashire man born into a farming family in the Fylde district but had broadened his experience while in the Royal Artillery in the First World War when he had managed a grass farm in India. During his time in India, like so many before him, he had contracted malaria from which he never fully recovered. Indeed he never went anywhere without his necessary pillbox full of quinine tablets. Unfortunately during his absence the family business had fallen on hard times and left Robert with no alternative but to seek employment wherever it could be found.

He saw an advertisement for the position of Secretary with the Peterborough Society, applied for the job, much to his surprise was invited for an interview and even more to his surprise, gained the job in preference to nearly 200 others. He quickly got to work. He had found his niche.

Perhaps the best way of summing up the great work he did and the standing in which he was held is to make reference to some of the points made at the presentation given in 1946 to mark his 25 years of service to the Society. At this presentation was a large gathering, not only of friends but also officers and members of the Society and the Peterborough County Branch Branch of the National Farmers' Union because Robert was the last Secretary of the Society who also acted as Secretary of the local N.F.U. Branch. In addition, he was Honorary Secretary to the Fitzwilliam Point to Point Races and, at that time, of the Local Wheat Committee.

While it is the custom to utter eulogies at such events, the tenor of what was said and indeed the whole atmosphere was indicative of the especially great respect and affection with which he was regarded. He was described as a 'valued counsellor' to all those from the farming community who came to him for advice which he always gave freely and wisely. The meeting was reminded how he had played such a prominent role in building up the size and standing of the Summer Show that the Secretaries of other Agricultural Societies came to him for advice. Someone even described him as the 'Prince of Show Secretaries'. Humorously, one speaker referred to his great expertise as a form-filler. Apparently, he could complete the forms that left others totally baffled.

Robert responded to the speeches with humour and dignity. As a northerner, he had been brought up to believe that the best people on Earth were those born north of the Trent but no sooner had he moved to Peterborough than he found himself among friendly people with whom it was a pleasure to work. He paid tribute to the current officers of the Society who, in his opinion, were well able to continue the excellent work of their predecessors. Not the least of the pleasures he had derived from his 25 years in harness was to see the rise in the membership of the local branch of the NFU. Every year membership figures had risen and, with typical Lancashire dry wit, he told the meeting that the Society now even had members in Yorkshire!

He congratulated the organisers of the presentation on the secrecy with which they had tried to surround the arrangements and to great laughter, he added that despite their clandestine efforts, it had been at least three weeks before he found out exactly what they were trying to do!

In accepting a cheque for £770, he said that he felt as if he had 550 friends for that was the number of subscribers who had contributed to the presentation.

After his death, an appeal raised £1,260 to help provide funds for the creation of a permanent memorial to his life and his work. As a result, the Robert Bibby Memorial Clock was erected over the ornamental entrance to the Eastfield Showground. This handsome tribute to a highly regarded man was unveiled by HRH Princess Alice, Duchess of Gloucester on the 16 July, 1958, during her visit on the second day of that year's Annual Show. Later the clock was removed from the Eastfield site and for some years stood on a column at the front of the Alwalton Showground close to the Oundle Road. In due course it was removed and repositioned yet again on the Administration Block at the Showground. In 1998 a descriptive plaque was erected on the side of the building informing visitors to the Showground of the history of this clock which commemorates a man who, although not born locally, the City made its own.

Roy Bird, M.B.E.

Born and brought up in the Fenland town of March, not far from Peterborough, Roy Bird, had a legal training and then went into the Army, serving with the British Army of the Rhine. On being de-mobbed in 1948, he joined the Peterborough Agricultural Society as Assistant to the then Secretary, Robert Bibby. Assisting Mr Bibby's work with the National Farmers' Union and the NFU Mutual Insurance Society, the Fitzwilliam Point to Point and the Local Wheat Committee, he gained a thorough and invaluable grounding in every aspect of the theory and practice of agricultural administration.

Roy continued working with Robert Bibby until the latter's death in 1957 when he was appointed Secretary of the Peterborough Agricultural Society in March of that year. In 1963 he additionally became Secretary of the Shire Horse Society and in 1965 took on the Secretaryship of the Peterborough Royal Foxhound Show Society, a position he holds to this day.

The East of England Show was held for the first time in 1968 and the East of England Agricultural Society itself was established out of a series of mergers between the Peterborough and various neighbouring Agricultural Societies in 1970. Roy became the first Secretary and Chief Executive of the East of England Agricultural Society and this without question immediately became one of the country's largest, most influential and progressive Agricultural Societies. He was awarded his M.B.E. in 1979 for services to agriculture. He retired in March 1992 by which time he had been deeply involved with the organisation of over 40 Annual Shows.

On 5 March 1992, the East of England Agricultural Society held its AGM followed by a Dinner attended by more than 180 guests, including many Past Presidents and Chairmen of the Society and its predecessor, the Peterborough Agricultural Society. His Royal Highness The Duke of Gloucester presented Roy with a very fine painting executed in 1991 by the well-known artist from Deeping St James, Derek Abel, which depicted the Main Ring at the Show at Alwalton in that year. The Show itself had been attended by Her Majesty The Queen and HRH The Duke of Edinburgh, the presence of such august visitors was a fitting finale to Roy's dedicated service as the lynchpin of the organisation and presentation of the Annual Show which had so widened its appeal on all fronts over the preceding years.

Joe Odam, a past Chairman and Deputy President, who was presiding at the Dinner and who had played a major part in organising this very seemly event, then gave a speech in which he paid a handsome tribute to Roy's achievements over the previous 44 years. Having worked closely with Roy for much of that time, he thought that the negotiations around the acquisition and planning of the Alwalton Showground and the later discussions and negotiations that brought together the various Agricultural Societies with their proud history of past achievement to form the East of England Agricultural Society, marked the undoubted peak of Roy's career and were achievements of immense importance to agriculture, not only locally but regionally. He also paid tribute to Roy's great skill in getting people to give time and effort voluntarily and willingly.

Roy recently stated with great modesty how grateful he had always been to have enjoyed the wholehearted, loyal, dedicated support and friendship of all Council members of the various societies which came together to make up the East of England Agricultural Society, the Stewards and particularly the members of both the Office and the Showground staff to whom he would wish to extend his gratitude. He emphasised that no Agricultural Society nor its Show Secretary could ever hope to make a success of the job without the enormous enthusiasm and co-operation which he found was so willingly given by those associated with the Society, no matter in what capacity.

Roy is married and he and his wife Norma enjoy sailing, golf and gardening. They still live in Peterborough.

CHAPTER TEN

Visions of the Future

This chapter consists of the contributions of a number of people eminent in the world of agriculture or its related areas and records their opinions and visions concerning the future of British Agriculture and agricultural shows as we approach the year 2000.

1. Sir David Naish
President, National Farmers' Union 1991- to 1997

I believe that amid all the changes that are taking place in our industry, it is vital that agricultural societies and shows can still play an important part in both leading and reflecting these changes. The world is shrinking in our ability to reach far-flung destinations either by travel or increasingly, through the speed of technological communication. But there will always be a job to do in explaining to an ever-curious public exactly how their food is produced and the role of the farmer not only in that process but also in being able to carry out the duty of care to the countryside.

As producers and providers of safe, high quality food, we must listen to our customers and to consumers. At the same time we must use every resource possible to promote the cause and objectives of this great industry of ours. There is a danger that the increasing use of technology as a tool of communication could divorce us from personal contact on a day-to-day basis. But the general public - as well as farmers and growers - will still welcome the chance to leave the confines of their offices or homes to gather at events such as agricultural shows.

Therefore, we have a fine opportunity for agricultural societies to use these shows and similar events to bridge the gap between town and countryside - by listening, by learning but above all by explaining. I sincerely hope, as we run up to the Millennium, that agricultural societies and shows can look to the future with renewed confidence knowing that they are a vital part of the agricultural scene.

2. J. Malcolm Stansfield, Director, Farm Management Unit, Department of Agriculture, The University of Reading

There is little doubt that over the years, even centuries, Show Societies have made a major contribution to the highly impressive development of British farming and the countryside as we see it today.

Some, although perhaps the minority, continue to be successful and have confidence in the future; whilst others are already at, or know they are fast approaching the 'cross roads' and have deep concerns as to the optimum route ahead.

Developing the 'best' future strategy for a Show Society is a very similar task to undertaking one for any business of farm. Objectives need to be discussed and agreed, but their achievement will only be possible if those objectives are realistic and take into account the physical and financial resources available, the market to be targeted and the capability of the people - especially the management!

The wealthier societies, with funds obtained from profitable trading or asset appreciation, are able to assist the local, regional or even national farming community by undertaking a range of activities such as funding research, staging technical events and conferences. The show (of whatever type) is still the key event for most societies. Change is inevitable and usually beneficial. Those who depend upon attracting the public to 'balance the books' do have to meet the social needs of their visitors. They can however also make an invaluable contribution to our industry by educating the visitors - if communication and interpretation methods are appropriate.

Formerly, it was the case that for success, it did not matter what was done, so long as it was done well. Now societies have to be undertaking the most appropriate market researched activity and operate with excellence, be it in the car parks, restaurants, conference hall or cattle lines.

3. Sir Richard Greenbury
Chairman and Chief Executive Marks & Spencer plc.

Marks & Spencer are proud of their close relationship with British Agriculture and as a Company have worked hard over many years to establish direct links with livestock producers and growers throughout the country. Food retailing is a fast moving, highly competitive sector and the supply chain has to adapt to meet the demands of the market place. It is therefore important that Agricultural Societies and Agricultural Shows demonstrate an awareness of current issues and respect the views of the consumer.

We believe that Agricultural Shows offer a unique opportunity for retailers to meet farmers and growers in a friendly and compatible environment. The Shows also provide an excellent opportunity to demonstrate to the general public, the progress farmers are making in terms of food safety, animal welfare and improved quality standards.

I know from our own suppliers, that certain aspects of British Agriculture are truly world class and I believe that Agricultural Shows provide an excellent opportunity to demonstrate these world class standards to the consumer. By playing their role in this way, Agricultural Shows can help further to improve the confidence of the consumer in British Food, which can only benefit British farmers in the long run.

Marks & Spencer will continue to give Agricultural Societies and Shows our support. We believe in the future of British Agriculture and will continue to work closely with our suppliers to help them achieve success.

4. E. Andrew Davies
Chairman 1998-9, East of England Agricultural Society

Reading the early chapters of this book, I was amazed how often history repeats itself. The number of times that Governments have thrown farmers to the mercy of world markets in the hope that they will increase productivity is quite astounding. As we approach the next round of World Trade Talks, British farmers are under threat of the end of all production-linked subsidies and a complete return to an open World Market.

The role of Agricultural Societies will be as important as the day they were founded.

Societies will need to appreciate that the Show is no longer the best place to educate farmers, but a marvellous opportunity to promote the industry to its consumers.

We see a determined effort by many well organised minority groups to discredit the farming industry. Who better than the Agricultural Societies to unite the industry? At our Show we must provide a fair and balanced view in an interesting and attractive manner. To our members we must provide the information to enable them to talk with confidence about the problems that concern our consumers. None of this willl be possible if we cease staging an Agricultural Show.

Technical, focused working demonstrations provide the ideal environment for our farmer members to gain new knowledge relevant to their business. They also have the added benefit of being extremely attractive and stimulating to our young members, which is proving a most useful way of starting an active involvement with the Society.

The young are our future and no effort must be spared to cultivate their interest and enthusiasm.

5. Mike Baunton President, Perkins Engines Company Limited

The relationship between Perkins Engines and the Agricultural Societies of England has been a long and very fruitful one. Even prior to 1932, when the Company was founded by Frank Perkins in Peterborough, Frank Perkins' forefathers were manufacturing agricultural machinery under the Barford Perkins name and were displaying their products at the shows of the old Peterborough Agricultural Society.

Frank Perkins always thought that the diesel engine was the power unit of the future because of its superior fuel efficiency. He believed that he could be the first to develop diesel engines which would deliver comparable performance to petrol equivalents. In 1932 he set about proving this to the world by establishing his own company, F. Perkins Ltd, in Peterborough, the forerunner of today's Perkins Engines which is now the world's leading suppliers of off-highway diesel engines.

Perkins' first success was with the 'Leopard' engine which was offered by Ford as an option in its Fordson tractors. More and more agricultural customers were signed up and encouraged to convert their machinery to take the broadening range of Perkins engines. Agricultural shows proved to be an important shop window for the company and attendance at these shows was an important part of spreading the message. Today, Perkins is one of the world's largest suppliers of diesel engines for the agricultural market. In the tractor section alone, the company produces the engines for around a quarter of all tractors manufactured throughout the world.

Agricultural shows have been instrumental in assisting the growth of the market for diesel engines. The strong base developed in the early days, initially in England, supported the global growth of the Company into all those markets where tractors are built or sold.

In the many years of Perkins participation at shows, they have been used to promote new and innovative developments of the engines specifically designed with the needs of the farmer in mind. As a result, diesel powered machinery will continue to dominate the world's agricultural markets."

6. Bill Samuel
Chief Executive, Peterborough City Council.

The relationship between the Agricultural Society and the City Council extends over a century. The City Council's earliest direct connection with agriculture came in 1876 with the purchase of the Rights to hold Fairs and Markets from the Ecclesiastical Commissioners. The right to hold the Cattle Market, which had been sold to a private company, together with the land, was acquired by the Council in 1891.

The Council were also farmers owning the 300 acre Sewage Farm adjoining the North Bank and the foreman, Edward Newton, regularly exhibited at the Show in the 1890s. Compared to the City's modest farm, the Soke of Peterborough County Council's small holdings amounting to some 3500 acres were quite considerable and as a new Unitary Council, the City Council is once again an owner and manager of farm estates. Both former Councils had amongst their serving members a fair proportion of farmers, dairymen, butchers and corn merchants. Some served on both Councils, like William Barford, who was not only Chairman of the Peterborough Agricultural Society for the last ten years of his life (1888-1898) but also a City Councillor and Alderman (1874-1898) and County Alderman (1889-1898). He was also responsible for bringing the Hound Show to Peterborough in 1887/8 where it has continued annually.

The Society, its activities and particularly the annual East of England Agricultural Show with its origins from the Peterborough Show at the site of the present Regional College, makes an impact on the City of Peterborough in many ways. The Show itself provides a showcase nationally for the city, its rural crafts and its local and regional agricultural prowess. It is a major tourist attraction and the event, and the interest shown in it, has a direct impact on the city's local economy.

The Society, the Council and the City have enjoyed a mutually supportive relationship both through land and property interests and the endeavours of local people. This relationship continues today and given the importance of the rural economy to the well being, health and prosperity of the City, it is a relationship that must continue to grow and be supported as we move towards the next millennium.

Authors' Comments

These six contributions emphasise and confirm the long-standing and continuing role played by Agricultural Societies in benefiting the farming fraternity in the widest sense by bringing them up-to-date information on the latest initiatives, new forms of technology and emerging ideas for improving output, productivity and profitability. The contributors' comments all indicate their confidence that this role will remain a valid one.

They also make clear the extent to which these organisations reflect the continual change of the wide social, economic, political and cultural circumstances in which they operate and the manner in which they can in turn help to develop mutual understanding between the farming community on the one hand and the general public, as consumers of home-grown food and users of the countryside, on the other.

The contributors recognise the integral part that the Peterborough Agricultural Society and its successor the East of England Agricultural Society have played and are continuing to play in linking the region's agricultural community with the citizens of Peterborough and district and their elected representatives in Local Government. Indeed there have been many local political dignitaries who have also served with distinction as officers of the Societies and have participated in their various committees. This is all part of the important contribution made by these groups to the prestige and prosperity of the City and it is no exaggeration to say that their activities form a crucial part of the history of the City of Peterborough area and the surrounding area.

Any history of Peterborough and its environs that did not include a full recognition of the crucial part played in local life by the two Societies would indeed be incomplete. It is the authors' hope that the publication of this present work constitutes a useful and informative contribution to a part of the City's fascinating history which has previously been seriously neglected by researchers.

APPENDIX ONE

Peterborough Agricultural Society 1874 List of Subscribers

	£	s.	d.
Addy, M.B., Market Deeping		10	0
Aldwinkle, Jun., Uffington, Stamford		10	0
Allatt, George, Marholm		10	0
Allatt, William, Dogsthorpe		10	0
Alderman, J., Thorney		10	0
Almond, J., Upton		10	0
Amies, Thomas, Peterborough		10	0
Andrews, John, Deeping, High Bridge		10	0
Argles, Rev. Canon, Barnack		10	0
Arnold, J., Thrapstone		10	0
Attenborough, R., Sawtry	1	00	0
Aveling, Jas.T., Eldernell		10	0
Baker, George, Barnwell		10	0
Baley, R.C., Thorney		10	0
Baley, R.W.,Thorney Dyke		10	0
Ball and Son, Rothwell		10	0
Bays, J.S., Peterborough		10	0
Barford & Perkins, Messrs, ditto	3	00	0
Bedford, Duke of, His Grace	20	00	0
Bedford, Miss, Thorney		10	0
Beeby, A., Peterborough		10	0
Beecroft,J., Eye		10	0
Beharrel, John, Thorney		10	0
Beharrel, Thomas, Eye		10	0
Bellars, William, Helpston		10	0
Berridge, John, Careby		10	0
Berridge, W., Burley-on-the-Hill		10	0
Bingham,T., Peterborough		10	0
Bird, Herbert, Fletton		10	0
Bird, John, Yaxley	1	00	0
Bird, John, Jun., Farcet	1	00	0
Bird, Rev. J., Peterborough		10	0
Blacker, Rev. S., Marholm		10	0
Blackman, John, Yaxley		10	0
Bland, R., Baston		10	0
Blott, J.T., Warmington		10	0

	£.	s.	d.
Blankley, Alfred, Grantham		10	0
Blunt, B., Whittlesey		10	0
Bodger, William, Peterborough		10	0
Bonser, James, Fotheringhay		10	0
Bonser, T., Elton		10	0
Boon, Wm., Walpole St Peters		10	0
Bower, Mrs., Orton		10	0
Bowman, Major, Duddington		10	0
Boyer, George, Marholm		10	0
Bradford, B., Farcet		10	0
Brainsby, T., Peterborough		10	0
Bright, Richard, Peterborough		10	0
Bristow, George, Peterborough		10	0
Bristow, James, ditto		10	0
Brown F., Peterborough		10	0
Brown, J., Willow Hall		10	0
Brown, Messrs., Wothorpe		10	0
Buck, John, Thorney		10	0
Bullimore,--, Deeping		10	0
Buckle, J.W., Peterborough		10	0
Burnham, H., Whittlesey		10	0
Burnham, J., ditto		10	0
Cadge & Colman, Peterborough		10	0
Cadwell, T., Peterborough		10	0
Callow, Thomas, Castor		10	0
Cape, Rev. W., Peterborough	1	00	0
Cautley, Rev. J., Thorney		10	0
Carpenter, L., St Neots		10	0
Carr, Edward, Whittlesey		10	0
Carter, Thomas, Ailsworth		10	0
Carter, W., Fletton		10	0
Carysfort, Rt, Hon the Earl of, Elton Hall	2	00	0
Catling,Capt., Needham Hall, Wisbeach		10	0
Cattel, J., Peterborough		10	0

	£	s.	d.
Chamberlain, J.H., Lyndon, Uppingham		10	0
Chambers, C., Orton		10	0
Chapman, G., Careby, Stamford		10	0
Chapman, M., Newborough		10	0
Chapman, W., Peterborough		10	0
Cheney W.J., Gidding Grange		10	0
Chesham, Lord	5	00	0
Clarabut, E., Peterborough		10	0
Clarabut, W., ditto		10	0
Clarke, J.S., ditto		10	0
Close, Thomas, Barnack		10	0
Close, W., Collyweston		10	0
Cole, James, Werrington		10	0
Cole, John, Whittlesey		10	0
Cole, Wm., Northborough		10	0
Coles F., Holme		10	0
Coles, Mrs., ditto		10	0
Collingwood, W., Peterborough		10	0
Cooke, H., Postland		10	0
Cooke, W.R., Moulton		10	0
Cooke, Rev J., Chesterton		10	0
Cooke, W., Yaxley		10	0
Cooper, Mrs., South Witham, Grantham		10	0
Copley, George, Peterborough		10	0
Core, John, ditto		10	0
Core, William, Fletton		10	0
Cory, Rev. R., Stanground		10	0
Cotchin, J., Peterborough		10	0
Cotterill, G., Eye		10	0
Craig, John, Fotheringhay		10	0
Crane, W,W., Thorney		10	0
Cranfield, W., Buckden		10	0
Crawley, S., Thorney		10	0
Crick, Charles, Peterborough		10	0
Crosbie, J., Holme		10	0
Cunningham J.A., Peterborough	1	00	0
Cutlack, C., Peterborough		10	0
Daintree, C., Fenton		10	0
Dainty, D., Belmisthorpe		10	0
Dainty, Geo., Easton, Stamford		10	0
Dainty, W., West Hay, Wansford		10	0
Dandy, R., Peterborough		10	0
Davis, A., ditto		10	0
Deacon, L.J., Esq., ditto		10	0
Deacon, Samuel, Polebrook		10	0
Dickens, J.T., Barnack		10	0
Dods, Richard, Barholm		10	0
Durrant, P, Oundle		10	0
Dyball, J., Bedwell Park, Hatfield	1	0	0

	£	s.	d.
Earl, S., Morborne		10	0
Edgson, Mrs., J.W., Elton		10	0
Edwardes, C.B. Esq.,	2	0	0
Edis, Isaac, Dogsthorpe		10	0
Edwards, J., Buckworth		10	0
Edwards, Thomas, Eye		10	0
Egar, S., Thorney		10	0
Ellis and Everard, Peterborough	1	00	0
Elsey, Henry, ditto			
Elston, J.T, Thorney			
English, H.H., Esq, Peterborough	1	0	
Everitt, J.H., Thornhaugh		10	0
Exeter, the Marquess of, Burleigh	2	0	0
Exley, Chas., Peterborough		10	0
Faux, E., Yaxley		10	0
Faux, R., ditto		10	0
Fellowes, E.,Esq., M.P., Ramsey	2	0	0
Fisher, A.G., Peterborough	1	0	0
Fitzjohn,W., Whittlesey		10	0
Fitzwilliam, Hon.C.W., M.P., Alwalton	2	0	0
Fitzwilliam, Hon.G.W., Milton	15	00	0
Fletcher, Howard, Newborough		10	0
Fletcher, Joshua, Jr., Postland		10	0
Flint, Wm., Washingley		10	0
Fowler, J and Co., Leeds	1	0	0
Fox, John, Peterborough		10	0
Freeman, J., Southwick		10	0
Freeman, J., Glapthorne		10	0
Fullard, George, Thorney		10	0
Fullard, Thos., ditto		10	0
Fullard, William, ditto		10	0
Fuller, A., Ramsey		10	0
Fyson, A., Warboys		10	0
Gaches, W.D., Esq., Peterborough		10	0
Gale, S., Kelmarsh		10	0
Gates, H.P., Esq., Peterborough	2	0	0
Gaunt, George, Peterborough		10	0
Gee, Edward, Wisbech Fen		10	0
Gee, George, Deeping Fen		10	0
Gery, W.H.W., Norman Cross House	1	0	0
Gibbs, David, Thorney		10	0
Giles, W.B., Glinton	1	0	0
Glew, Alfred, Peterborough		10	0
Godfrey, W., Tansor		10	0
Golden, C., Benwick, Ramsey		10	0
Golden, Thomas, Cambridge		10	0
Golden,J., Plantation farm, Ramsey		10	0
Golden, S., St Mary's Ramsey		10	0
Gollings, R., Woodstone		10	0

	£	s.	d.
Goodliffe, J., Connington		10	0
Goodman, A., Sen., Willow Hall		10	0
Goodman, A., Jnr., ditto		10	0
Goodyer, H., Cawthorpe		10	0
Goodyer, William, Peterborough		10	0
Gordon, F., Thornhaugh		10	0
Grant, Thos., Whittlesey		10	0
Graves, J., Peterborough		10	0
Gray, Rev. E., Alwalton		10	0
Gray, Geo., Uffington Grange		10	0
Green, Benj., Harringworth		10	0
Green, George, Alconbury		10	0
Greenwood, John Burnley		10	0
Gregory, T., Duddington		10	0
Gregory, W.H. Jun., Apethorpe		10	0
Gresswell, J.G., Peterborough		10	0
Grieves, J., Orton		10	0
Griffin, C.W., Werrington	1	0	0
Griffin, John, Peterborough		10	0
Griffin, R.H., Boro' Fen		10	0
Griffin, W.E., Werrington		10	0
Griffin, P., Eye		10	0
Grounds, B.W., Jun., Whittlesey		10	0
Grounds, Daniel. Eldernell		10	0
Hack, S., Postland		10	0
Hall, H.J., Peterborough		10	0
Hall, James, ditto		10	0
Hanger,T.(Exrs)., Willow Hall		10	0
Haynes, W., Deeping St James		10	0
Hardy, Thomas, Cloot House, Crowland		10	0
Harris, W., Orton		10	0
Harrison, G., Peterborough		10	0
Hayes and Son, ditto		10	0
Heanley, M., ditto		10	0
Heanley, Wm., ditto		10	0
Heathcote, J.M., Esq., Connington	2	0	0
Hemment, John, Thorney		10	0
Heyes, J., Warmington		10	0
Hibbet, J. Deeping St James		10	0
Hill, Charles, Dogsthorpe		10	0
Hill, George, Edenham		10	0
Hill, George, Peterborough		10	0
Hill, H.A., Elton		10	0
Hill, James, Oundle		10	0
Hill, Thomas, Peterborough		10	0
Hind, Jos., Orton		10	0
Hockney, G., Wittering		10	0
Horrell, H., Thorney		10	0
Holdich, Yorke, Elton		10	0
Holland, W., Market Deeping		10	0

	£	s.	d.
Hope, W., Wellingborough		10	0
Hopkinson, John, Dogsthorpe		10	0
Hopkinson, W., Sutton		10	0
Hopper, R. North Bank		10	0
Hornsby, J., Castle Gate, Grantham		10	0
Hornsby, Philip, Grantham		10	0
Horrell, Wm., Thorney		10	0
Horrell, H., ditto		10	0
House, John, Peterborough		10	0
How, James, Broughton		10	0
Huddlestone, Robert, Thorney		10	0
Hughes, E., Barnack		10	0
Hughes, G., Glassmore, Whittlesey		10	0
Hunt, Edward, Barnack		10	0
Hunt, the Rt. Hon. G.W., M.P., Wadenhoe	2	0	0
Huntley, the Most Noble Marquis of	2	0	0
Hurry, W.J., Thorney		10	0
Hurry, S., Jun., ditto		10	0
Ireson, Chas., Peterborough		10	0
Isaac, H., North Bank		10	0
Jackson, J.L., Esq., Bainton House		10	0
Jakes, John, Peterborough		10	0
Johnson, Edward, Helpston		10	0
Johnson, Edward, Maxey		10	0
Johnson, Thomas, Walton		10	0
Jones, Henry, Littleport		10	0
Jones, Richard, Woodstone		10	0
Jones, T.H., White Fen, Benwick		10	0
Jones, W., Coates, Whittlesey		10	0
Jones, W., New Fletton		10	0
Kesteven, the Rt. Hon. Lord, Casewick	2	00	0
King, Wm., North Luffenham		10	0
Kirby, Mrs., Glapthorn		10	0
Lamb,H., Norwood House, March		10	0
Laurence, J., Esq., Elton	1	00	0
Laurence, J., Jun., Esq., Warmington		10	0
Laxton, John, Morborne		10	0
Layton, John, G.N.R. Hotel	1	00	0
Leeds, A., Eyebury		10	0
Lenton, W., Oundle		10	0
Lilley, E.J., Peterborough		10	0
Little, George, Farcet		10	0
Little, H., Peterborough	1	00	0
Little, J.E., Peterborough		10	0
Little, J.J., Singlesole		10	0
Little, W., Littleport		10	0

	£	s.	d.
Livett, Z., Esq., Whittlesey		10	0
Lock, J., Nassington		10	0
Longland, J., Grendon		10	0
Lotan, H., Oundle		10	0
Lound & Son, Peterborough	1	00	0
Lowe & Betts, Stamford	1	00	0
Mackenzie, J., Peterborough		10	0
Mackinder, J., ditto		10	0
Mackinder, R., ditto		10	0
Mann, W., Marholm		10	0
Mann, George, ditto		10	0
Marriott, J.T., Castor		10	0
Marshall & Sons, Peterborough	1	00	0
Martin, S., Helpston		10	0
Matthews, Louis, Dogsthorpe		10	0
Mattley, Miss, Peterborough		10	0
Mawer, E., Thorney		10	0
Maxwell, G.S., Walton		10	0
Maxwell, James, Thorney		10	0
Maxwell, R.W., Orton		10	0
Maxwell, W.G., Walton	1	0	0
Mayo, S., Peterborough	3	00	0
Middleton, S., Waternewton	1	00	0
Midland Banking Co., Peterborough	1	00	0
Mills, Thos., Esq., Fletton	2	00	0
Molecey, J.M., West Deeping		10	0
Moore, John, Peterborough	1	00	0
Moore, W.G., Eye		10	0
Morley, G., (Exors), Ryhall		10	0
Morris, B.W., Thorney		10	0
Mossop, Rev. C., Etton		10	0
Murfitt, Edward, March		10	0
Musson, W., Peterborough		10	0
Neal, James, Deeping St. Nicholas		10	0
Neal, K., Langtoft		10	0
Needham, F., Woodnewton		10	0
Newbon, J., Peterborough		10	0
Newton, C., Alconbury Weston		10	0
Newton, J.A., Fotheringhay		10	0
Noble, F., Peterborough		10	0
Nottingham, Thos., Ufford		10	0
Odam, W., (Exors), Walton		10	0
Odam, M., ditto		10	0
Odam, John, Belsize		10	0
Odam, J., Jnr., Newborough		10	0
Paley, W., Esq., M.D., Peterborough	1	00	0
Palmer, J.E., ditto		10	0

	£	s.	d.
Pank, John, Oxney		10	0
Parker, Thomas, Dogsthorpe		10	0
Parnell, Thomas, Thorney		10	0
Parnell, William, ditto		10	0
Pate, H.T., Ely		10	0
Pawlett, Ed., Deeping St. James		10	0
Paviour, W., Peterborough		10	0
Pearce, W., Helpston		10	0
Pearson, J.H., Peterborough		10	0
Peck, Robert, Barton Fen		10	0
Percival, A., Esq., ditto		10	0
Percival, George, ditto		10	0
Percival, John, Esq., ditto		10	0
Percival, Thomas, Wansford	1	00	0
Percival, Thos., Jun., ditto	1	00	0
Perkins, S., Thorney		10	0
Peterboro', Rt. Rev. Lord Bishop of	2	00	0
Phipps, P & R., Peterborough	1	00	0
Pilkington, George, Baston		10	0
Poles, John, Alwalton		10	0
Porter, W.G., Esq., Peterborough		10	0
Pratt, Revd., Canon, ditto		10	0
Prior, B., Leighton		10	0
Provost, Mrs., French Drove, Thorney		10	0
Richardson, T.W., Fletton		10	0
Richardson,W.,Peterborough		10	0
Roach, John, Gretton		10	0
Roberts, Chas., Peterborough		10	0
Roberts, W., ditto		10	0
Rose, J.D., Holme Fen		10	0
Rose, Saml., Great Ravely		10	0
Row, J.L., Woodston	1	00	0
Rowell, G., Longthorpe		10	0
Rowell, John, Bury, Ramsey		10	0
Ruddle, James, Peterborough		10	0
Sampson, C.J., Water Newton Mills		10	0
Sams, Mrs., Washingley		10	0
Santall, F., Peterborough		10	0
Savidge, James, Stamford		10	0
Savill, Geo., Ingthorpe, Stamford	1	00	0
Sawyer, James, Peterborough		10	0
Sawyer, J.S., Whittlesey		10	0
Schoeppler, L., Peterborough		10	0
Scoley, Thomas, ditto		10	0
Scotney, R., Thorney		10	0
Searson, R., Deeping Fen		10	0
Searson, S., Peterborough		10	0
Seargeant, E.B., ditto		10	0

	£	s.	d.
Searson, -, Tallington		10	0
Searle. T., Horsey, Whittlesey		10	0
Sewter, Thomas, Werrington		10	0
Sharpe, J.J., Roughton, Kettering		10	0
Sharpe, T.D., Peterborough		10	0
Sherard, Lord, Glatton	2	00	0
Shelton. T., Oundle		10	0
Shepperson. Lewis, Whittlesey		10	0
Shrive, E., Peterborough		10	0
Sisman, W., Buckworth		10	0
Skrimshire, E.A., Esq., Stanground		10	0
Slater, John, Orton		10	0
Smith, G.S., Shive Farm, Stamford		10	0
Smith, J.H., Ailsworth		10	0
Smith, J.T., Old Fletton		10	0
Smith, Sidney, Newark		10	0
Smith, Thomas, Newark		10	0
Smith, William, Glinton		10	0
Smith, J., Kings Delph, Whittlesey		10	0
Southam, G., Esq., M.B., Peterborough		10	0
Southam, H., Yarwell		10	0
Southam, G.W., Thorney		10	0
Southampton, Rt. Hon. Lord,	2	00	0
Southwell, J.B., Benwick		10	0
Southwell, Thomas, Stilton		10	0
Speechley, James, Yaxley		10	0
Speechley, Thomas, ditto		10	0
Speed, Charles, Exton		10	0
Spencer, R., Keystone		10	0
Staffurth, E.H., Whittlesey		10	0
Stanley, S., Peterborough		10	0
Staplee,W., Oxney	1	00	0
Stapleton, G., Market Deeping		10	0
Stokes, J.H., North Luffenham		10	0
Stokes, Samuel, Duddington		10	0
Stokes, Thomas, Glaston		10	0
Stokes, Thos., Warmington		10	0
Stokes, Thos., Caldecote, Rockingham		10	0
Strickson, W., Thurning		10	0
Strong, C.I. Esq., Thorpe Hall	2	00	0
Sturton & Son, Peterborough		10	0
Symonds, G., Westwood		10	0
Tate, H.J., Holme		10	0
Tebbs, Charles, Peterborough		10	0
Tebbitt, J., Doddington		10	0
Templeman, J.H., Carlby		10	0
Thompson, Mrs. A., Peterborough		10	0
Thompson, J., Peterborough		10	0
Thompson, W.J., Crowland		10	0
Thurstun, J., Peterborough		10	0
Tiptaft, C., Tinwell Lodge		10	0
Tiptaft, W.C., Stamford		10	0
Todd, Henry, Northborough		10	0
Todd, Henry, Jnr, Peterborough		10	0
Tomlinson, Jas., Lutton Marsh	1	00	0
Toon, W., Woodwalton		10	0
Traylen, J., Chesterton		10	0
Triplow, W., Chatteris		10	0
Truman, J.H., Whittlesey		10	0
Turner, George, Eastfield		10	0
Turner, J., Haddon Grange	1	00	0
Turnill, H., Sawtry		10	0
Turnill, J., Sawtry		10	0
Twelvetree, J., Eye		10	0
Ullett, W., Higney Grange		10	0
Underwood, J.W., Warmington		10	0
Vawser, Edward, Holme		10	0
Vergette, E., Peterborough		10	0
Vergette, E., Newborough		10	0
Vergette, F., Peterborough		10	0
Vergette, G.G., Deeping		10	0
Vergette, J.B., Holme	1	00	0
Vergette, J.T., Whittlesey		10	0
Vergette, R., Eldernell		10	0
Vergette, R.P., Boro' Fen		10	0
Vergette, Samuel, Glinton		10	0
Vergette, Thomas, Boro' Fen		10	0
Vergette, T.H., ditto		10	0
Vergette, W., Peterborough		10	0
Wadsworth, C., Peterborough		10	0
Wagstaff, Thomas, Chesterton		10	0
Wagstaff, W.J., Stilton		10	0
Wagstaff, T.H., Elton		10	0
Walden, Thomas, Northborough		10	0
Walker, R., Longthorpe	1	00	0
Walker, T.,Esq., M.D., Peterborough		10	0
Walker, T.J., M.D., Peterborough		10	0
Walker, E., M.D., Peterborough		10	0
Wallis, John, Thorney		10	0
Wallis, R., ditto		10	0
Walton, J.A., Burleigh		10	0
Ward, D., Ashton, Stamford		10	0
Warth, M., Sutton, Ely		10	0
Warwick, E., Thorpe		10	0
Warwick, J., Thorpe		10	0
Warwick, R.B., Stanground		10	0
Watson, F., Linwood Farm, March		10	0
Watson, Charles, March		10	0

	£	s.	d.
Watson, Rev., Water Newton		10	0
Watts, W., Junior, Leicester		10	0
Wattam, F., Oakham		10	0
Webster, D., Nunton		10	0
Webster, J., Peakirk	1	00	0
Webster, R., Peakirk		10	0
Wells, W., Esq., Holme Wood	3	00	0
Wenlock, T., Peterborough		10	0
West, Samuel, Upwell		10	0
Whalley, G.H., M.P., Ruabon, N. Wales	2	00	0
Whitehead, S., Glatton		10	0
Whitehead, W., Etton		10	0
Whitehead, W., Woollaston		10	0
Whitney, W.S., Glatton		10	0
Whitting, W., Esq., Thorney	2	00	0
Whittome, A., Whittlesey		10	0
Whittome, C., Jun., Doddington	2	00	0
Whitwell, H., Peterborough		10	0
Whitwell, J., Peterborough	2	00	0
Wiggington, James, Thorney		10	0
Wilcox, Thomas, Langtoft		10	0
Wilders, J.L., Casterton Magna		10	0

	£	s.	d.
Willis, James, Farcet,		10	0
Wilson, S., Peterborough		10	0
Winder, E., Clapton Hall		10	0
Wiseman, H., Holbeach		10	0
Windower, C., Huntingdon		10	0
Wood, R., Clapton		10	0
Wood, Thomas, Denton		10	0
Wood, Thomas, Carlby		10	0
Woods, W., Edith Weston		10	0
Wooll, W., Manor Farm, Hartford		10	0
Wortley, E., Ridlington		10	0
Wortley, Wm, Newborough		10	0
Wrensfordsley, H. Esq., Temple, London	2	00	0
Wright, Charles, Wansford		10	0
Wright, Robert, Tallington		10	0
Wright, T.P., Helpston		10	0
Wyman, G., Esq., Peterborough		10	0
Wyman, W., Fletton		10	0
Wyman, H., Connington		10	0
Yeoman, J., Esq., Milton	2	00	0
Yeathird, S.J., Sawtry		10	0

APPENDIX TWO

1797-1967
Presidents of
The Peterborough Agricultural Society

1968 onwards
Presidents of
The East of England Agricultural Society

(With Deputies where appropriate)

1797-1833	The Rt Hon. The Earl Fitzwilliam.	1893	Henry Wickham, Esq.
1834-1857	Earl Charles William Fitzwilliam.	1894	The Duke of Bedford.
1858-1868		1895	Sir Walter Gilbey, Bart.
& 1871	Hon. George W. Fitzwilliam.	1896	The Marquess of Exeter.
1869	Hon. C.W. Fitzwilliam, M.P.	1897	P.A. Muntz, Esq., M.P.
1870	Hon. C.W. Fitzwilliam, M.P.	1898	Viscount Milton, M.P.
1872	The Marquis of Huntly.	1899	Sir Gilbert Greenall, Bart.
1873	The Hon. G.W. Fitzwilliam.	1900	V.C.W.Cavendish, Esq. M.P.
1874	The Hon. C.W. Fitzwilliam, M.P.	1901	Capt. W.H. Duncombe.
1875	The Marquis of Huntly.	1902	The Marquess of Exeter.
1876	The Earl of Carysfort.	1903	Lord Lilford.
1877	Wm. Wells, Esq.	1904	The Hon. N. Chas. Rothschild.
1878	The Marquess of Exeter.	1905	The Hon. A.E. Fellowes, M.P.
1879	The Marquis of Tavistock.	1906	H.L.C. Brassey, Esq.
1880	The Earl Fitzwilliam.	1907	G.C.W. Fitzwilliam, Esq., M.F.H.
1881	The Marquis of Huntly.	1908	Jas. Hornby, Esq.
1882	Lord Burghley, M.P.	1909	F.E. Muntz, Esq.
1883	The Earl of Lonsdale.	1910	Rev. W. Hopkinson, M.A.
1884	Lord Chesham.	1911	Lord Rothschild.
1885	Wm. Wells, Esq.	1912	The Duke of Devonshire.
1886	Lord Wantage.	1913	Lord Lilford.
1887	H.P. Gates, Esq.	1914	The Marquess of Exeter.
1888	G.C.W. Fitzwilliam, Esq.	1919	Stanley Brotherhood, Esq.
1889	The Hon. J.W. Fitzwilliam, M.P.	1920	Stanley Brotherhood, Esq.
1890	Major G. Wickham.	1921	The Rt. Hon. Lord Lilford.
1891	Lord Rothschild.	1922	Lt.Colonel. D. Davies, M.P. F.M.H.
1892	The Hon. A.E. Fellowes, M.P.	1923	Earl Fitzwilliam, M.F.H.

1924	Sir Leonard Brassey, Bart, M.P.
1925	The Rt. Hon. Lord Lilford.
1926	Major J.A. Morrison, D.S.O.
1927	Brig. Gen. W. Strong, C.B., C.M.G.
1928	Col. The Hon. F.V. Willey, M.F.H.
1929	Lord Burghley.
1930	J.A.. de Rothschild, Esq., M.P.
1931	Lt-Col. Sir A Weigall, K.C.M.G.
1932	H.R.H. The Duke of York.
1933	The Duke of Beaufort, M.F.H.
1934	Lt. Col. F.G.D. Colman, M.F.H.
1935	Major Gordon Foster.
1936	T.W. Fitzwilliam, Esq., M.F.H.
1937	Brig.Gen. A.F.H.Ferguson.
1938	H.R.H. The Duke of Gloucester, K.G.
1939	The Marquess of Exeter, K.G.
1946	Lord Brassey of Apethorpe.
1947	H.R.H. The Duke of Gloucester, K.G.
1948	His Grace The Duke of Portland, K.G.
1949	His Grace The Duke of Buccleuch G.C.V.O.
1950	H.R.H. The Princess Elizabeth, Duchess of Edinburgh.
1951	Lord de Ramsey, M.F.H.
1952	Major Sir Richard Proby, Bart.
1953	The Earl Fitzwilliam, M.F.H.
1954	The Lady Ethel Wickham.
1955	The Rt. Hon. The Earl Spencer.
1956	The Marquess of Exeter, K.G.
1957	H.T.L. Young, Esq.
1958	Major the Hon. G.W. ff.Dawnay , M.C.
1959	W. Geoffrey Barford, Esq.
1960	H.R.H. The Duke of Gloucester, K.G.
1961	F.A. Perkins, Esq.
1962	The Marquess of Exeter, K.C.M.G., M.F.H.
1963	Rupert V. Cobb, Esq.
1964	W.S. Smith, Esq., O.B.E.
1965	The Earl Fitzwilliam.
1966	The Earl Fitzwilliam.
1967	The Rt. Hon. The Marquess of Exeter.
1968	The Rt.Hon, The Lord De Ramsey, KBE, TD, DL.
1969	The Rt.Hon Lord Luke, KCVO, JP, TD, DL.
1970	HRH The Prince Philip, Duke of Edinburgh. KG, KT.
	The Lord Hemingford (Deputy).
1971	The Lord Netherthorpe, LL.D, B.Sc.
1972	HRH Princess Alice, Duchess of Gloucester.
	Col. G.T. Hurrell, OBE, K. St.J. (Deputy).
1973	Sir Henry Plumb, DL, MEP.
1974	Lt. Col The Hon. Peter E. Brassey.
1975	W.T. Godber, Esq., CBE.
1976	His Grace The Duke of Buccleuch and Queensberry, KT.
1977	Col. G.T. Hurrell, OBE, K.St.J.
1978	Finn Olav Gundelach, Esq.
1979	The Rt. Hon. The Earl Ferrers.
1980	HRH The Duke of Gloucester, GCVO.
	Joseph Odam, Esq., JP. (Deputy).
1981	The Countess Fitzwilliam.
1982	Sir Richard Butler.
1983	Poul Dalsager, Esq.
1984	Sir Peter Proby, Bt.
1985	The Rt. Hon. The Lord Boardman, MC, TD, DL.
1986	The Rt. Hon Robin Leigh-Pemberton, PC.
1987	Frans H.J.J. Andriessen, Esq.
	Major Peter Bletsoe Brown, TD, DL. (Deputy).
1988	Sir John Harvey-Jones, MBE.
1989	HRH. The Princess Royal, GCVO.
	The Hon. Lady Hastings, M.Phil, MFH. (Deputy).
1990	The Earl of Dalkeith, DL.
1991	S.C. Whitbread, Esq.
1992	Sir John Quinton.
1993	The Rt. Hon. The Lord Pym, PC, MC, DL.
1994	The Rt. Rev. William Westwood, The Lord Bishop of Peterborough
1995	Edmund Vestey, Esq., MFH, DL.
1996	Sir John L. Lowther, KCVO, CBE, JP.
1997	Her Majesty The Queen.
	HRH The Duchess of Gloucester GCVO (Deputy).
1998	Lady Victoria Leatham, D.L.

APPENDIX 3

Peterborough Royal Foxhound Show Society Presidents from 1878

1878-81	The Earl Fitzwilliam	1923	HRH The Prince of Wales
1882	The Earl Spencer, KG	1924	John C. Straker, Esq., MFH
1883	The Earl of Zetland, MFH	1925	The Duke of Northumberland, MFH
1884	The Earl Ferrers, MFH	1926	Sir F. Villiers Forster, Bt., MFH
1885	The Earl of March	1927	Lt Col. Sir Dennis Boles, Bt. MFH
1886	Lord Willoughby De Broke, MFH	1928	The Earl Bathurst, MFH
1887	The Earl of Yarborough	1929	Edward E. Barclay, Esq., MFH
1888	G.Lane Fox, Esq. MFH	1930	Douglas Crossman, Esq., MFH
1889	Colonel Anstruther Thomson	1931	Capt. Esme Arkwright, MFH
1890	A. Brassey, Esq. MFH	1932	HRH The Duke of York
1891	The Marquis of Huntly	1933	The Duke of Beaufort, MFH
1892	R. Chandos-Pole, Esq., MFH	1934	The Earl of Lindsay, MFH
1893	The Earl of Lonsdale	1935	Lewis Priestman, Esq., MFH
1894	Thomas Parrington, Esq.	1936	Capt. F.B. Atkinson, MFH
1895	HRH The Prince of Wales, KG	1937	The Duke of Buccleuch, MFH
1896	Sir W. Williams-Wynn, Bt. MFH	1938	HRH The Duke of Gloucester, KG.
1897	Lord Middleton, MFH	1939	The Earl Fitzwilliam, MFH.
1898	G.C.W. Fitzwilliam, Esq., MFH	1946	Capt. T.W. Fitzwilliam, MFH
1899	The Earl Bathurst, MFH	1947	Lord Poltimore, MFH
1900	The Duke of Beaufort, MFH	1948	Lt. Col. Sir Harold Nutting, Bt.
1901	The Duke of Leeds, MFH	1949	Lord Knutsford
1902	The Duke of Sutherland, MFH	1950	Col. J.G. Lowther, MFH
1903	Lord Ribblesdale	1951	Lord Irwin, MFH
1904	Lord Annaly, MFH	1952	Col. R. Milvine, MFH
1905	The Earl Manvers, MFH	1953	Lt Col. Sir P. Farquhar, Bt. MFH
1906	The Viscount Portman, MFH	1954	Maj. M.E. Barclay, MFH
1907	Arthur Wilson, Esq.	1955	George Evans, Esq.
1908	W. H. Wharton, Esq., MFH	1956	Maj. J.G. Morrison, MFH
1909	The Earl Fitzwilliam, MFH.	1957	The Duke of Northumberland, MFH
1910	Lord Willoughby De Brooke, MFH	1958	The Earl of Feversham, MFH
1911	The Marquis of Zetland	1959	The Earl of Yarborough, MFH
1912	Lord Leconfield, MFH	1960	Capt. G.E. Bellville
1913	Lord Ribblesdale	1961	Lt Col Sir W. Williams-Wynn Bt MFH
1914	Sir W. Williams-Wynn, Bt. MFH	1962	The Duke of Beaufort, KG. MFH
1920	The Duke of Beaufort, MFH	1963	The Marquess of Exeter, MFH
1921	Lord Willoughby De Brooke, MFH	1964	Col. Sir Ralph S. Clarke, MFH
1922	Lt. Col. David Davies, MFH	1965	The Duke of Beaufort, KG. MFH

1966	The Earl Fitzwilliam, MFH.
1967	Sir Alfred L. Goodson, Bt. MFH
1968	G. Percival Williams, Esq.
1969	Rt. Hon The Earl of Halifax, MFH
1970	The Rt. Hon The Lord Ashton of Hyde
1971	Capt. R.E. Wallace, MFH
1972	The Lord Daresbury, MFH
1973	The Earl of Dalkeith, MFH
1974	M.F. Burry, Esq
1975	HRH The Princess Anne
1976	Sir R. Buchanan Jardine Bt. MFH
1977	The Countess of Feversham, MFH
1978	HRH The Prince of Wales
1979	Marcus Kimball, Esq.
1980	Lt. Col. G.A. Milway-Smith, MFH
1981	Capt. C.G.E. Barclay, MFH
1982	R.J.G. Berkeley, Esq., MFH
1983	H.R. Goschen, Esq., MFH
1984	Maj. G.A. Gundry, MFH
1985	Mrs. Ulrica Murray Smith
1986	His Grace The Duke of Northumberland
1987	Lt. Col. Sir Henry Tate, Bt
1988	The Earl of Yarborough, MFH
1989	Lady Crossman
1990-91	His Grace The Duke of Beaufort
1992	Sir W. Williams-Wynn, Bt, MFH
1993	E.H. Vestey, Esq. DL MFH
1994	Maj. J.A. Warre, MC
1995	His Grace The Duke of Beaufort
1996	Lt. Col Sir John Miller, GCVO DSO MC
1997	The Hon. Lady Hastings, DL. M.Phil. MFH
	(Deputy) Sir Stephen Hastings, MC MFH
1998	Sir Ralph Carr-Ellison, T.D. Ed.

Bibliography

Much use has been made of the archives of the *Lincoln, Rutland and Stamford Mercury*, the *Peterborough Advertiser* and the *Peterborough Standard.*

Armstrong, A. — *Farm workers: A Social and Economic History 1770-1980,* London 1988

Chambers, D. & Mingay, G.E. — *The Agricultural Revolution 1750-1880,* London 1966

Creasey, J.S. & Ward, S.B. — *The Countryside Between the Wars 1918-40,* London 1984

Ernle, Lord — *English Farming Past and Present* (1st Ed,1912), 6th Ed, London1961

Gibson, M.L. — *Aviation in Northamptonshire: An Illustrated History,* Northampton 1982

Goddard, N. — *The Royal Agricultural Society of England, 1839-1988,* London 1988

Grigg, D. — *English Agriculture: An Historical Perspective,* Oxford 1989

Grigg, D. — *The Agricultural Revolution in South Lincolnshire*, London 1966

Hall, V. — *A History of the Yorkshire Agricultural Society, 1837-1987*, Leeds 1988

Hudson, K. — *Patriotism with Profit: British Agricultural Societies in the Eighteenth and Nineteenth Centuries,* London 1972

Kerridge, E. — *The Agricultural Revolution*, London 1967

Mingay, G.E. — *Rural Life in Victorian England,* Stroud 1990

Mingay, G.E. (Ed) — *The Agrarian History of England and Wales (Vol 6),* London 1989

Newby, H. — *Country Life: A Social History of Rural England*, London 1987

Orwin, C.S. — *A History of English Farming,* London 1949

Orwin, C.S. & Whetham, E. — *A History of British Agriculture 1846-1914,* London 1964

Perren, R. — *Agriculture in Depression, 1870-1940,* Cambridge 1995

Ruddock, J.G. — *The Lincolnshire Agricultural Society: Its predecessors and development*, Lincoln 1983

Tebbs, H.F. — *Peterborough,* Cambridge 1979

Thompson, F.M.L. — *English Landed Society in the Nineteenth Century,* London 1963

Watson, J.A.S. — *History of the Royal Agricultural Society of England,* London n.d. *1839-1939*

Whetham, E.H. — *British Farming, 1939-1949,* London 1952

Index

Page numbers in italics refer to illustrations.
Streets are in Peterborough unless stated otherwise.